SPACEX FROM THE G

5th EDITION

By Chris Prophet

DEDICATION

To fellow sci-fi writer Elon Musk: -

I use mere words where
Elon writes with steel and aluminum

Elon Musk presents SpaceX Falcon 9 v1.0 Rocket

CONTENTS

Introduction: SpaceX Paradigm

There's a storm gathering in aerospace, vast potential accumulating in the clouds, spreading threads of lightning to everything it touches. At its center is a company called SpaceX, comparatively small by aerospace standards yet holding a position of enormous moment. Their business model is a new paradigm for creating advanced products, which could potentially revolutionize commerce - even how we think about the nature of business. SpaceX punch well above their weight and are enormously ambitious. Their corporate mission is to colonize other planets(1), which surprisingly they seem capable of achieving, given their focus, sacrifice and financial resources. The first goal is to colonize Mars, not in a generation or two but in less than a decade(2).

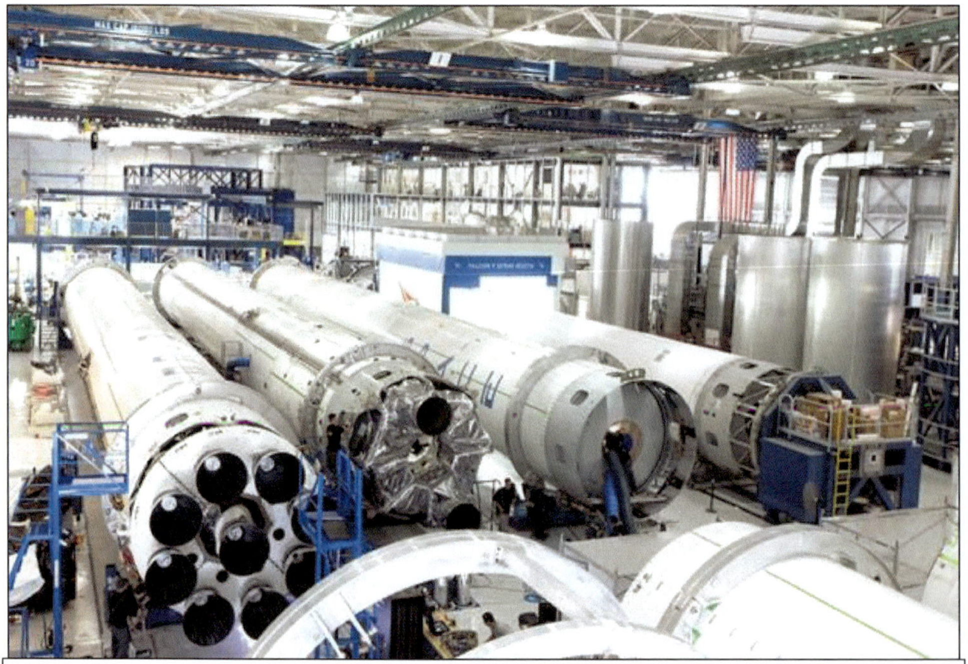

Falcon 9 assembly line at SpaceX California

Some might argue any attempt to colonize space is a distraction, possibly futile, and won't benefit the way they live. However, everything SpaceX does on their way to Mars is new, so they will need to develop some entirely distinct and creative technologies. Needless to say these new technologies and techniques could profoundly affect how we live on Earth. More importantly, whatever societal structure SpaceX decide to adopt on the red planet could have inordinate

"SpaceX was founded under the belief that a future where humanity is out exploring the stars is fundamentally more exciting than one where we are not. Today SpaceX is actively developing the technologies to make this possible, with the ultimate goal of enabling human life on Mars." ~ *SpaceX (careers) website.*

influence on the evolution of our own society. If SpaceX demonstrate a new and viable model for how to live more effective and fulfilled lives on a frontier planet like Mars – it will inevitably draw us, influencing everything from our political system down to our personal relations. When Mars becomes self-sufficient, the world turns.

Mars could have inordinate influence on human development

To conclude, we are privileged to live in a time when truly great changes are wrought, as humanity rapidly evolves into a space faring civilization. Evolution on this scale requires enormous drive and endurance and the name on that engine of change is: -

SPACEX

[1]http://www.spacex.com/about
[2]https://youtu.be/vbC_jT2mIdY?t=1873

Chapter 1: How SpaceX Was Born

Mortal Dream

The year is 2000, dawn of the new millennium. Elon Musk is approaching a fork in the road; one path descends to darkest fate, the other rises to shining destiny.

Elon is incredibly wealthy for someone still in his twenties, after receiving $165m from the sale of PayPal, which he helped to create. Having a surplus of time on his hands and an embarrassment of riches he quite reasonably decided to go on a long overdue holiday to Brazil and his native South Africa. Unfortunately, upon his return to the U.S. he was struck down by a deadly strain of cerebral malaria. He came incredibly close to death and was only saved by the last minute intervention of a malaria specialist doctor.

Saturn V moon rocket

This near-death experience had a chastening effect on Elon, with all his riches and talent, what had he actually achieved? He perceived the world hadn't made the huge advances he'd hoped for as a boy. In some cases technological progress, like the advances made in rocketry during the Apollo program, seemed to be receding.

During his college years he identified five things which would most influence our future: the internet, rewriting human genetics, sustainable energy, artificial intelligence and human space exploration(1). Space now seemed accessible (given his new-found fortune), so he decided to move to Los Angeles, which had long been a hub for aerospace development. He taught himself the basics through reading old rocketry text-books and technical manuals then reached out to fellow space enthusiasts and industry professionals. His friend Robert Zubrin (President of the Mars Society) suggested they should send a group of mice to LEO (Low Earth Orbit) to discover the long-term effects of low gravity, including reproduction.

However, Elon Musk wanted to reinvigorate space exploration and make humans multi-planetary. He reasoned this would require something more impactful, like sending a breeding group of mice to Mars and back, all paid for out of his own pocket. Elon hoped this grand gesture would help demonstrate the next logical step of settling Mars, even raising a family there, was feasible with current technical understanding. After discussing this project with friends, he received a giant wheel of Gruyère cheese from Jeff Skoll, along with the advice: "The fornicating mice would need a lot of cheese(2)."

Undaunted Elon arranged a number of informal gatherings with his space industry contacts to help validate the idea. At some point this plan morphed into something slightly more practical, yet still capable of capturing the world's imagination.

Mars Oasis Project

Deployable Greenhouse on Mars

Following a series of meetings, they conceived what they really needed to launch was a "Mars Oasis." A privately funded project could land a miniature greenhouse on Mars to demonstrate life was possible on that distant and seemingly inhospitable planet. In essence, this graphic demonstration was designed to rally public support and spur NASA funding for the exploration of Mars.

Elon Musk assembled a team of friends and space industry experts (including his college friend and fellow internet entrepreneur Adeo Ressi, NASA/DOD space specialist Jim Cantrell, ex-CIA technologist Mike Griffin) and together they investigated how to best execute this bold endeavor. They discovered the main stumbling block to this mission would be the high cost to launch payloads into space, particularly if they used a domestic (i.e. US) rocket. The least expensive option was to use an old USSR rocket called Dnepr-1; a modified SS-18 ICBM which had been decommissioned as part of the START Treaty. However, negotiations didn't go well in Moscow, caused in large part by the asking price for the rocket. A price of $21m was agreed for three Dnepr-1 launch vehicles but when the money was produced, the Russian negotiators revealed that sum would only buy one rocket...

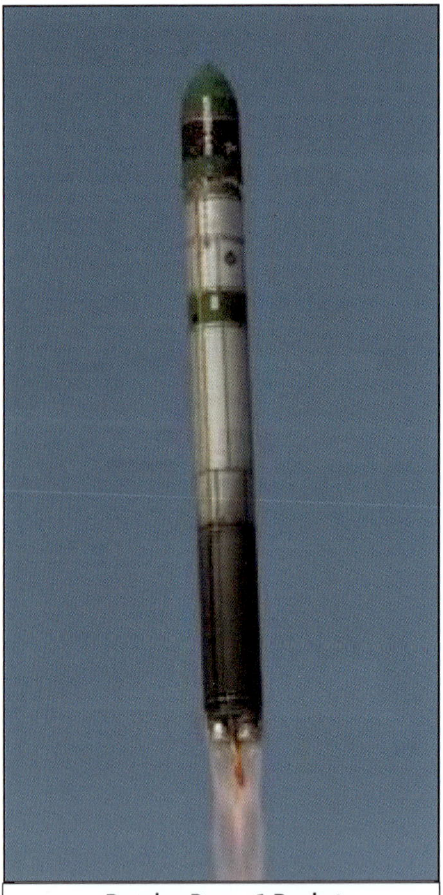

Russian Dnepr-1 Rocket

"Then we started having meetings with the Russian space program, which is basically fueled by vodka. We'd all go in this little room and every single person had his own bottle in front of him. They'd toast every two minutes, which means twenty or thirty toasts an hour. 'To space!' 'To America!' 'To America in space!' I finally looked over at Elon and Jim and they were passed out on the table. Then I passed out myself. They had two more trips scheduled to Russia; now Ressi decided, as he says, 'I didn't like dealing with Russians,' and told Musk he wasn't going back. Musk went anyway."

"On the second trip, Musk brought his wife, Justine – 'I think that's the trip when the lead Russian designer started spitting at us,' Cantrell says – and on the third and final trip he brought his money. He was ready to buy three Russian ICBMs for $21 million when the Russians told him that no, they meant $21 million for *one*. 'They taunted him', Cantrell says. They said, 'Oh, little boy, you don't have the *money*?' I said, 'Well, that's that'(3)."
~ *Esquire Magazine*

Elon Musk stormed out of negotiations - but on the return flight home made a shocking announcement: "I think we can build a rocket ourselves." Then he proceeded to show them some spreadsheets he had been working on, on his laptop, which indicated they could independently build a commercially viable launch vehicle.

I Think We Can Build a Rocket Ourselves

Elon reasoned the material costs to actually build a rocket were only 3% of the final launch price. From a purely economic perspective, the main impediment to space exploration was the cost of the actual launch vehicle. Consequently, if they built their own rocket, using modern manufacturing techniques and materials, they could open an extremely worthwhile and viable company. He further reasoned this private company could grant them sufficient revenue and space launch experience to attempt the holy grail of space exploration: the human colonization of Mars.

Following a great deal more research and consultation, Space Exploration Technologies Corporation (SpaceX) was formed in Los Angeles California, June 2002. The company was explicitly set up to spearhead man's exploration of Mars and generate increased space activity through lowering launch costs. They have come a long way since 2002, considering their inordinate financial success and currently hold an order book worth more than $12 billion(4), you might easily imagine their more altruistic goals, like opening space to everyone, could become more soft focus. However, SpaceX remains in private hands, in fact it is majority owned by their own enthusiastic employees, which allows them to stay laser focused on their goal of reaching Mars.

Mars or Bust

"When decisions are being made, people will often bring it up: 'is this going to work for the Mars mission?' That question is always considered when decisions are being made; Mars doesn't always win, but that concern is always examined(5)." ~ *Robert Rose (former Director of Flight Software at SpaceX)*

It should be noted this Mars-centric attitude derives from the top and is hallmark Elon Musk behavior. He relentlessly researches and pursues an idea until it becomes reality or he determines a better way to achieve his overarching goal. Admittedly Elon Musk is a driven idealist but he has chosen to champion our long term survival. He believes that humanity is far too exposed living all in one place. For it to have any future we must find safe harbor beyond the sanctuary of Earth, amongst the wider planets and stars. More importantly, he is moving Heaven and Earth to make it happen.

"I think it [going to Mars] is the next natural step... One can also think of it from the standpoint of life insurance. There's some chance, either as a result of something humanity does, or as a result of something natural like a giant asteroid hitting us or something, that civilization - life as we know it - could be destroyed... I think it's also one of the most inspiring and interesting things that we could try to do. It's one of the greatest adventures that humanity could ever embark upon. You know, life has to be more than about solving problems. If all that life is about is solving problems then why bother getting up in the morning? There have to be things that inspire you that make you proud to be a member of humanity(6)." ~ *Elon Musk*

Awaking to new challenges and opportunities

Committing oneself to such an ambitious undertaking would hardly seem 'reasonable' – but if you want to achieve something truly momentous, sometimes you need unreasonable.

"The reasonable man adapts himself to the world; the unreasonable one persists in trying to adapt the world to himself. Therefore all progress depends on the unreasonable man."
~ George Bernard Shaw

CONCLUSION

Everything SpaceX does has a single unifying goal: truly All Roads Lead to Mars (ARLM).

[1]http://www.businessinsider.com/elon-musk-on-startalk-with-neil-degrasse-tyson-2015-4?r=US&IR=T
[2]ISBN: 9780062301239, "Elon Musk: Tesla, SpaceX, and the Quest for a Fantastic Future" by Ashlee Vance
[3]http://www.esquire.com/news-politics/a16681/elon-musk-interview-1212/
[4]https://twitter.com/pbdes/status/960930344588361728
[5]http://lwn.net/Articles/540368/
[6]https://youtu.be/xrVD3tcVWTY?t=729

Chapter 2: What SpaceX Have Achieved So Far

THE EARLY YEARS

Like most private startups SpaceX started small, occupying only a few rooms in a downtown Los Angeles hotel in June 2002. They quickly progressed to a 10,000 sqft warehouse in El Segundo California(1), where they commenced building their first two stage rocket called the Falcon 1(2), in homage to the Millennium Falcon from the Star Wars milieu. The Merlin engine also made its debut on this rocket; its pintle design possesses superior performance and is relatively inexpensive to build (pintle was first used on the Apollo program for the Lunar Module Lander). The first Falcon 1, powered by the first Merlin 1A engine, made its maiden flight in March '06.

Falcon 1 launch at Omelek Island

Unfortunately, this first flight failed, as did the second - and third. However, the third Falcon 1 flight, which used the upgraded Merlin 1C engine, made it to space but then barely failed, due to a staging issue. They discovered the more powerful Merlin 1C engine had provided residual thrust after shut-down, which

"Optimism, pessimism, fuck that; we're going to make it happen. As God is my bloody witness, I'm hell-bent on making it work." ~ *Elon Musk after Falcon 1 Flight 3 failure* caused the first and second stages to collide after stage separation. So if they simply allowed more time between main engine cut off and stage separation they would potentially had a winner. But with little money left Elon Musk knew he could only afford to build one more rocket. Fortunately, the fourth Falcon 1 launch managed to place a simulated payload (called 'RatSat') in orbit on September 28, 2008. When Falcon 1 flew again on July 14, 2009 they succeeded with their first commercial satellite launch, called RazakSAT, for the Malaysian government. This proved to be the last Falcon 1 flight as they swiftly transitioned to an even more capable two stage rocket called the Falcon 9(3).

Sunrise on Falcon 9 Rocket

THE 2008 CRASH

In 2008 SpaceX moved to its new headquarters, a one million sqft production facility, in Hawthorne California. The Falcon 9 rocket, whose first stage was powered by nine Merlin 1C engines, would need a lot more room to manufacture and SpaceX planned some really big things for the future. At that time the company had little launch revenue and largely subsisting on NASA development money from the COTS (Commercial Orbital Transportation Services) program and venture capital investment.

Unfortunately, this meant they were already financially stressed when the financial downturn struck at the end of 2008. SpaceX were developing Falcon 9 and the Dragon cargo spacecraft(4) in parallel and desperately needed more money to stay in business. CEO Elon Musk committed all his personal fortune to convince potential investors that SpaceX remained a worthwhile investment. He committed everything and even borrowed money from friends to pay for personal expenses, like rent for his accommodation.

As December drew to a close, SpaceX finally managed to secure the necessary funding from their investors. NASA too lent their support when they awarding SpaceX a $1.6 billion CRS (Commercial Resupply Service) contract to provide 12 cargo flights to the ISS(5). In a few days they had gone from potentially insolvent to a completely viable company with a solid long term future. From this point on SpaceX never looked back and continues to grow in leaps and bounds.

DAWN OF A NEW ERA

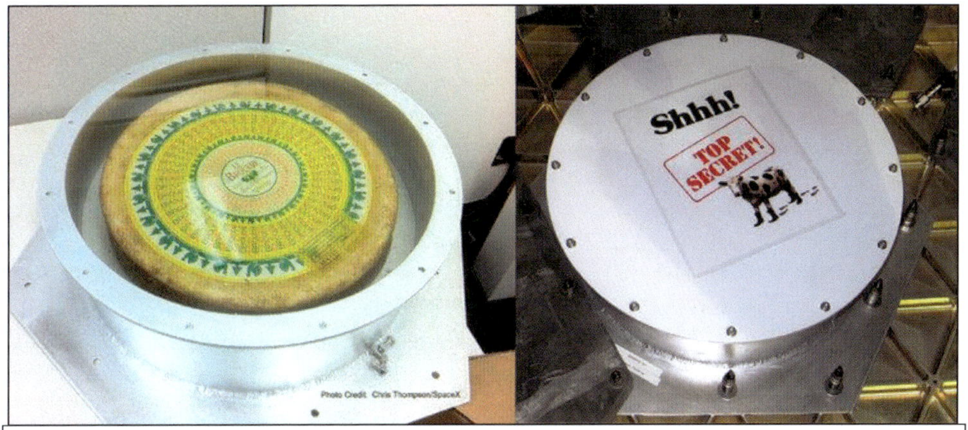

Top secret cargo and cargo container, credit SpaceX

SpaceX launched their first Falcon 9 rocket and Dragon spacecraft on June 4 2010, proving the design of their two stage rocket and capsule on the first attempt. During the next flight, called COTS-1 (the first Commercial Orbital Transport Service demonstration flight for NASA), they included a top secret cargo – a giant wheel of Gruyère cheese! Unfortunately this wasn't the same wheel of cheese Jeff Skoll jokingly gave to Elon so many years before but it was certainly a humorous riposte.

For the third Falcon 9 flight COTS-2+, they boldly decided to combine two missions, maneuvering close to and finally berthing with the ISS (International Space Station) on May 25, 2012.

Dragon arrives at International Space Station

This marked the first time a privately operated spacecraft had rendezvoused and berthed with another spacecraft, the first of firsts. SpaceX made good on their promise to NASA, providing many more cargo missions to the ISS and safely returning thousands of kilograms of cargo in the Dragon's pressurized cabin.

On April 18, 2011, NASA awarded SpaceX a CCDev2 contract(6) to develop the Dragon into a passenger carrying spacecraft. SpaceX plan to launch their crewed spacecraft, called Dragon 2, in early 2020 and no doubt score another first.

Elon Musk at the controls of SpaceX Dragon 2 spacecraft

The Falcon 9 rocket has been upgraded three times to date (Falcon 9 v1.1, Falcon 9 v1.2/Full Thrust and Falcon 9 Block 5), significantly improving its launch potential. So far, it has successfully sent payloads to Low Earth Orbit (LEO), Geostationary Transfer Orbit (GTO) and even helped send a Deep Space Climate Observatory (DISCOVR) to L1, a Sun-Earth Lagrange point(7).

REUSABLE ROCKETS

SpaceX never rest on their laurels and are always seeking to improve their existing launch technology. Since the first Falcon 1 they have strived to find a way to recover the expended stages of their rockets, which normally break up after staging. Early attempts to recover the first stage using parachutes proved unfeasible, so they decided to try an even more ambitious recovery procedure: landing their booster stages propulsively after each launch.

Propulsive Landing Modifications

Falcon 9 v1.1 in pre-flight prep, SLC-40 hangar, Cape Canaveral

They needed a lot more propellant to land propulsively, so the Falcon 9 was stretched by 13.5m to provide the additional propellant tank capacity. In addition retractable grid fins were mounted high on the booster body to provide aerodynamic guidance and four extendable legs fitted near the base of the rocket. A beefed up Reaction Control System (RCS) was also required, accompanied by an entire new suite of flight control instruments and avionics software to support the challenging regimen of acrobatic flight maneuvers required for return and safe landing.

Propulsive Landing Technique

After stage separation the Falcon 9 booster flips through 180 degrees and uses a three engine burn to boost back towards the launch site. As it approaches the upper atmosphere the grid fins deploy in an X-wing formation to provide additional drag and carefully guide re-entry. Before the booster reaches the dense lower atmosphere its descent speed is minimalized by another three engine burn. The stage then descends at terminal velocity towards the landing pad, using the grid fins to guide its final approach. Then as the ground rushes up to meet it, the booster simultaneously deploys four pneumatic landing legs and performs a final burn with its center engine to achieve a vertical landing (in what SpaceX call a hover-slam maneuver).

Stage recovery and reuse

The best part for me was when, as a team, we started to realise that landing was possible. We had all worked together to make a million individual breakthroughs and there was a moment when all these breakthroughs came together to show that, as long as everything on the rocket did what it was supposed to, we'd land. That was when it felt like everyone's work was eventually going to pay off, even though it still took us a few tries before it finally did(8)! ~ *Lars Blackmore, Principal Mars Landing Engineer for SpaceX.*

Falcon 9 (Flight 20) landing at LZ-1 Cape Canaveral

After many trials, SpaceX finally managed to achieve what had hitherto seemed impossible on December 21, 2015, at Landing Zone 1 (LZ-1) Cape Canaveral(9), when they successfully landed the OG2 booster (Falcon 9 flight 20). During 2017 they began to use these recovered boosters to send new Falcon 9 second stages and payloads to orbit. The Falcon 9 booster comprises 70% of the overall cost of the rocket(10) so if they can reuse a booster on multiple flights (with little or no refurbishment) then they have effectively managed to reduce the cost of space access by 70%. The cost to manufacture Falcon 9 is already a fraction of conventional launch providers, so by reusing boosters they have likely achieved a ten-fold reduction in launch cost. It's fair to say rocket reusability is an evolutionary leap in space technology which could herald a new era of space exploration.

"We want to open up space for humanity, and in order to do that, space must be affordable." ~ *Elon Musk*

Dawn of Falcon Heavy, LC39-A Cape Canaveral, credit SpaceX

Not content with developing their heavy lift Falcon 9, SpaceX needed even more launch power to achieve their dreams. So in early 2018 they hauled out to the launch pad their first SHL (Super Heavy Lift) rocket, called Falcon Heavy. This launch vehicle combines three Falcon 9 cores in one stack, with the center core topped by a standard F9 upper stage. This approach allowed them to minimize development cost (through utilizing existing Falcon 9 hardware, production machinery and test facilities) and maximize payload capacity, by staggered staging of the 3 boosters. Inspirationally, Elon Musk chose to use his own Tesla Roadster as the payload on this test flight, with a space suited mannequin behind the wheel called "Starman." Fortunately this test launch proved a success, sending the now famous Roadster and Starman beyond the orbit of Mars.

In a technical tour-de-force, SpaceX managed to recover both side boosters from this Falcon Heavy demo flight, which performed synchronized landings, descending side-by-side onto LZ-1 and LZ-2 at the Cape.

Falcon Heavy Side Boosters land at the Cape, credit SpaceX

The advent of Falcon Heavy cannot be overstated. SpaceX now have a rocket that can lift more than twice the payload of any existing launch vehicle, and even more impressively, it was all designed, built and financed exclusively by SpaceX.

"...we said, 'Nope [to government funding]! We just wanna build it, you can buy it when it's ready and we'll charge you for the service(11).'" ~ *Hans Koenigsmann, SpaceX Vice President of Build and Flight Reliability*

Falcon Family Comparison Table				
Max Payload	**LEO**	**GTO**	**Mars**	**Pluto**
Falcon 9	22.8 mt	8.3 mt	4.02 mt	n/a
Falcon Heavy	63.8 mt	26.7 mt	16.8 mt	3.5 mt

Falcon Heavy opens up new vistas of possibility, because it is now officially the most powerful rocket in operation. Classed as a 'Super Heavy Lift' (SHL) launch vehicle, it's capable of placing more than 50 metric tons into Low Earth Orbit (LEO). Effectively that means FH can launch any planned payload to any location in our solar system. Here's a sample of some already announced missions which could be handled by Falcon Heavy: -

- **Europa Clipper** – robotic reconnaissance of Europa, sixth moon of Jupiter
- **Psyche** – robotic mapping of metallic asteroid 16 Psyche in the main asteroid belt
- **WFIRST** – infra red telescope for exoplanet hunting and dark matter research

More significantly, Falcon Heavy can launch much heavier payloads to the outer solar system, allowing orbital/lander spacecraft to be sent, instead of more conventional flyby missions. Falcon Heavy marks the start of a new era for SpaceX, as they begin to transition into a fully-fledged "space transport company."

Elon's midnight cherry Roadster passes the moon on its way to Mars, with Starman at the wheel
credit SpaceX

CONCLUSION

SpaceX's drive to develop new spacecraft, rapid reusability and bigger launch vehicles might seem like random acts of commercial altruism but taken as a whole they demonstrate how SpaceX intend to pursue a much larger goal. These achievements are in fact stepping stones on the path to Mars - or as Elon puts it: "the Holy Grail objective(12)."

[1]https://www.youtube.com/watch?v=omBF1P2VhRI
[2]https://en.wikipedia.org/wiki/Falcon_1
[3]https://en.wikipedia.org/wiki/Falcon_9
[4]https://en.wikipedia.org/wiki/Dragon_(spacecraft)
[5]http://www.nasa.gov/home/hqnews/2008/dec/HQ_C08-069_ISS_Resupply.html
[6]https://en.wikipedia.org/wiki/Commercial_Crew_Development
[7]https://en.wikipedia.org/wiki/Deep_Space_Climate_Observatory
[8]http://www.eng.cam.ac.uk/news/alumni-stories-meet-principal-rocket-landing-engineer-spacex
[9]https://www.youtube.com/watch?v=1B6oiLNyKKI
[10]http://www.forbes.com/sites/alexknapp/2014/04/25/spacex-falcon-9-reusable-stage-landed-safely-in-the-atlantic/#4e44cc1c5402
[11]https://www.teslarati.com/spacex-executive-nobody-paid-us-to-make-falcon-heavy/
[12]https://youtu.be/afZTrfvB2AQ?t=1311

Chapter 3: Why SpaceX Wants to Settle Mars

SpaceX consider Mars the next big step in space exploration. Like anything they do, their decision to pursue Mars was logical, incisive and practical. It should be helpful to explore the reasoning behind this seminal decision, which should allow us some insight into their unique perspective on space exploration.

SPACE SETTLEMENT

In contemporary times, space exploration has become increasingly expensive, sporadic and some might argue sclerotic. But if there was someplace habitable in space, somewhere within easy reach, space travel would likely become commonplace. For example, if NASA had managed to establish a Moon base following the Apollo program, it would have likely spurred the space industry to find cheaper and more efficient ways to transport goods and people to that destination. These new technologies and techniques would have then enabled them to explore Mars, the asteroid belt, even the gas giants, until they finally reached the edge of the solar system. We know this is likely the case because that is what has occurred, in microcosm, with the ISS. When NASA decided to create a habitable destination in LEO it led SpaceX to develop some new and exceptionally low cost spacecraft to support it.

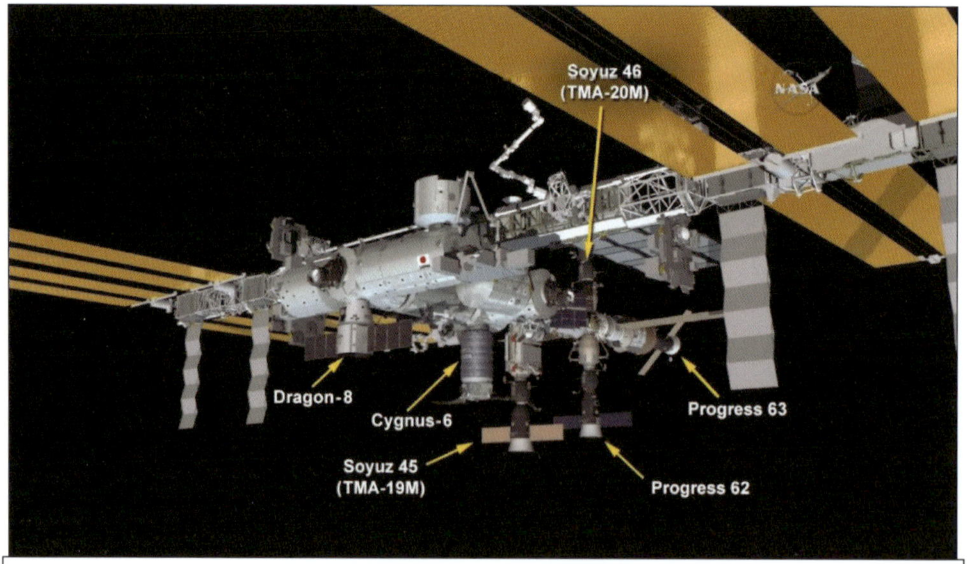

International Space Station with attendant spacecraft

Unfortunately the ISS is unlikely to be sustainable beyond 2028. What is required then is a fresh start in a new location, now LEO is becoming increasingly commercialized. The moon might appear the best prospect for establishing a new habitation due to its proximity to Earth. However, setting up a temporary habitation would provide only a tiny stimulus to the space industry compared to establishing a long-term settlement. The moon's extremely low gravity (roughly one sixth of Earth's) means anyone born there would likely possess physiological problems e.g. poor bone and muscle development, inadequate immune system etc. In addition, the moon has a twenty eight day rotational period (which makes agriculture difficult) and possesses no appreciable atmosphere.

Given these difficulties, planets would appear the best place to establish a sustained human settlement. Planets possess reasonable gravity (vital for long term habitation) abundant material resources (which should prove relatively simple to mine), some are even blessed with an atmosphere (to a greater or lesser degree). The outer planets are high gravity giants, which preclude any possibility of landing or long term survival. They possess habitable moons but overall are too far to travel (using present day technology) and receive insufficient sunlight for agriculture. Which leaves the inner planets (i.e. Mercury, Venus and Mars) as the most suitable candidates for long term habitation.

MERCURY SETTLEMENT

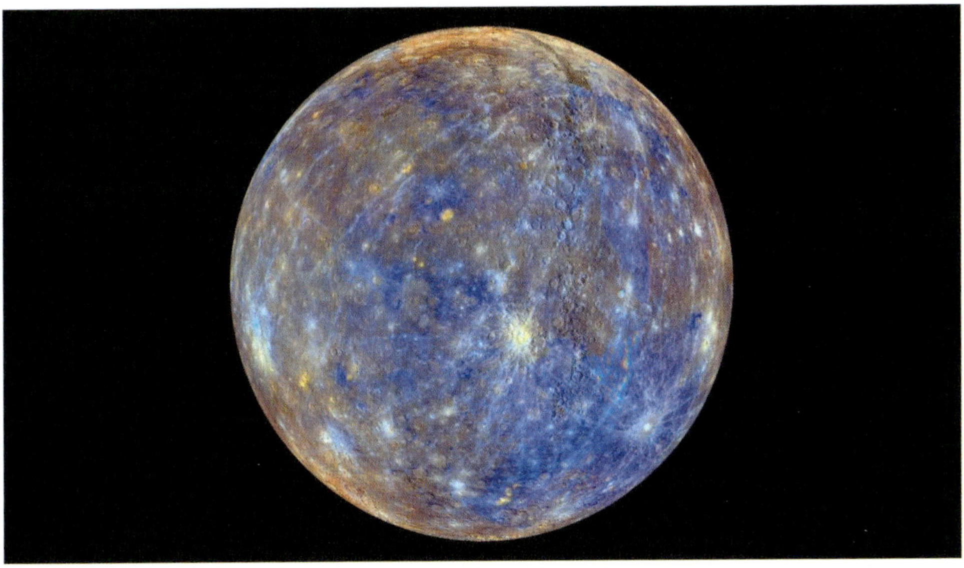

The planet Mercury is nearly tidally locked to the sun, which is to say it rotates excruciatingly slowly (it takes 58.6 Earth Days to complete one rotation). I say excruciating because Mercury is comparatively close to the sun, so days there are extremely hot - and tortuously long. Surface temperatures vary from +427°C (800°F) during the day to −173°C (−280°F) at night(1). It possesses no appreciable atmosphere and a relatively poor magnetosphere which means its surface receives large doses of solar and cosmic radiation. Surface gravity is 0.38 g (38% of Earth's) which is low but quite possibly adequate. Hence siting a sustained habitation at some shady spot at the pole might be feasible, although potentially parlous due to ambient radiation.

Mercury Settlement	
Pros	**Cons**
Adequate Gravity	Extreme Heat/Cold
Polar Settlements	No Direct Sunlight at Poles
	Radiation
	No Atmosphere

VENUS SETTLEMENT

The surface gravity on Venus is similar to Earth's (0.904 g) and it rotates once every 243 Earth days(2). Its atmosphere is extremely dense and consists mostly of carbon dioxide, which is a strong greenhouse gas. This results in a surface pressure 92 times greater than Earth's and surface temperatures of more than 462°C. Habitation might be possible at the cloud top level, using dirigibles or some equivalent suspension technology. However, Venus' weak magnetosphere provides very little protection from solar and cosmic radiation, which could in turn provoke violent electrical storms.

Venus Settlement	
Pros	**Cons**
Reasonable Gravity	Radiation
Cloud Top Settlements	Violent Storms
	Acidic Atmosphere

Venus Atmospheric Maneuverable Platform (VAMP)

Living in a cloud hopping settlement on Venus sounds like fun, your solar powered aircraft forever following the sun. Unfortunately such 'settlements' are unlikely to be self-sustaining and you could never land - with any luck...

MARS SETTLEMENT

Mars's surface gravity of 0.376 g(3) (roughly one third of Earth's) appears adequate, it has abundant mineral resources and even possesses a rudimentary atmosphere. In addition, the low gravity and thin atmosphere makes the red planet an ideal jumping off point for the nearby asteroid belt or outer planets. Mars lacks a coherent magnetosphere but it is more distant from the sun which reduces the effect of solar radiation. A solar day on Mars lasts for 24h 37m 22s and hence very similar to our own world. The average surface temperature is low at −55°C (compared to a balmy 15°C on Earth) but tolerable given the provision of adequate heat and shelter. Interestingly, in the distant past Mars appears to have experienced similar surface conditions to Earth, which implies it might possibly be terraformed. Various techniques could be used to turn back Mars's biological clock, imbuing it with a breathable atmosphere, warm oceans and stable biome. So, on balance, Mars would appear the best choice for long term habitation. In addition, its close proximity to the asteroid belt means it is strategically located at the crossroads for future space development. The asteroid belt promises to provide a cornucopia of raw materials for space expansion. These materials will be used to manufacture goods in space to service an entirely new sector of the economy: the space habitation and transport industry.

Mars Settlement	
Pros	**Cons**
Adequate Gravity	Radiation
Earth-like Day	Cold
Partial Atmosphere	
Superior Launch Capability	
Strategic Position	
Terraforming Potential	

CONCLUSIONS

1. Mars has all the resources needed to become self-sustaining thus the best place to found a space settlement.

2. SpaceX only have eyes for Mars. They're in it for the long haul – of humanity.

[1]https://en.wikipedia.org/wiki/Mercury_(planet)
[2]https://en.wikipedia.org/wiki/Venus
[3]https://en.wikipedia.org/wiki/Mars

Chapter 4: How SpaceX are Preparing for Mars

The best space agencies from the mightiest nations on Earth have attempted the arduous trek to Mars, using custom built satellites and robotic landers. Roughly a third of these landings have succeeded, proving it's unspeakably hard to achieve anything meaningful on the red planet, particularly in the long term. However, SpaceX was built from the ground up to transport people to Mars and have consequently evolved some extraordinary, some might argue extreme practices and technology, in order to reach this goal.

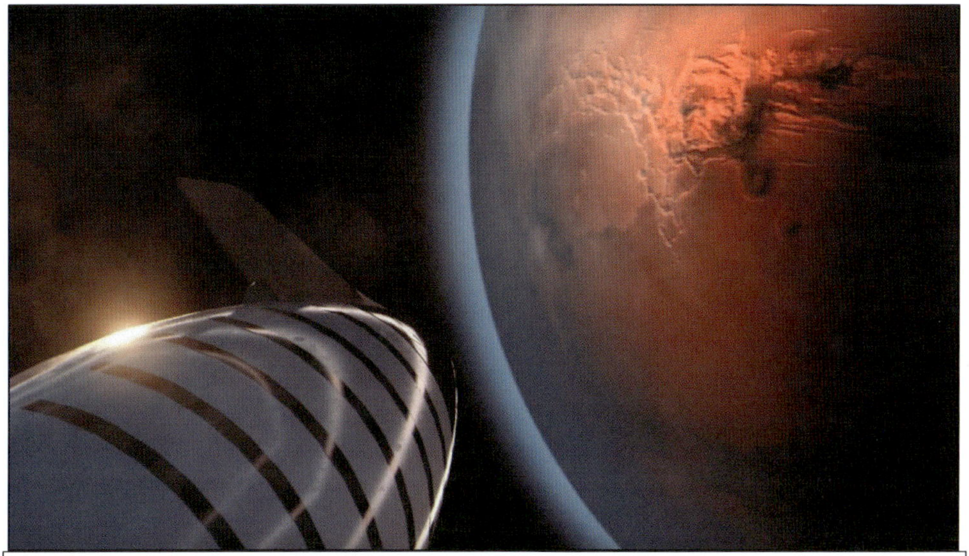

Colony vessel on approach to Mars

While these technologies and practices might grab the headlines, SpaceX's true strength stems from a fusion of many positive attributes, both subtle and extreme. So, let's explore some of the reasons why SpaceX have made themselves worthy of the challenge of Mars.

OWNERSHIP

SpaceX was originally established and currently owned by engineers. The company's CEO and CTO, Elon Musk, has co-founded two software companies (Zip2 and PayPal) and three major engineering companies: SpaceX, Tesla Motors and SolarCity. Similar to SpaceX, Tesla Motors and Solar City are both 'tech-heavy' companies with similarly grand ambitions. Tesla aims to help us transition away from hydrocarbon based transport to more environmentally sound electric vehicles.

SolarCity provides cheap solar power for private homes all the way up to government organizations. They intend to reduce or eliminate our dependence on less environmental power generation by revolutionizing the way we create and store electricity. On August 1, 2016, SolarCity was bought by Tesla, effectively creating a broad-based electrification company under the Tesla banner, to provide consumers with an environmental alternative to fossil fuels.

"Tesla was founded in 2003 by a group of engineers who wanted to prove that people didn't need to compromise to drive electric – that electric vehicles can be better, quicker and more fun to drive than gasoline cars. Today, Tesla builds not only all-electric vehicles but also infinitely scalable clean energy generation and storage products. Tesla believes the faster the world stops relying on fossil fuels and moves towards a zero-emission future, the better(1)."

So Elon Musk is not just a brilliant ideas man he is also a hardcore engineer. Given his grasp of emerging technology he is undoubtedly the right man at the right place and time.

SpaceX's President and COO, Gwynne Shotwell has a Master's of Science in Mechanical Engineering and an accomplished engineer who graduated from the automotive sector to aerospace. In fact every employee at SpaceX is a partial owner/private stock holder in the company through vested share options. This arrangement allows the engineers to make most of the decisions about how the company is run and the best engineering solutions to pursue. This strong engineer bias means the majority of the company's revenue is used to develop new rocket technology, test facilities and launch sites.

SpaceX HIF (Horizontal Integration Facility) at Pad 39-A, Cape Canaveral

SpaceX's corporate mission is to colonize Mars, so you could say these owner engineers own the problem.

"The path to the CEO's office should not be through the CFO's office, and it should not be through the marketing department. It needs to be through engineering and design(2)." ~ *Elon Musk*

RAW TALENT

"If you're wondering how SpaceX can do that [build and fly Starhopper in under a year], it's pretty simple. They hire the best talent. People want to work there because they do crazy things like build water towers and fly them 150 meters into the air on the way to Mars. Half a million people apply to the company a year(3)." ~ *Eric Berger, Senior Space Editor at Ars Technica*

Many young engineers view SpaceX as the apex for aerospace, even overtaking NASA as the most desirable employer in a 2019 student poll(4). For their part, SpaceX do much to foster this talent, offering many intern places for local high school students through to university graduates. For a multitude of reasons SpaceX attract the best talent possible, literally the cream of the crop. Essentially, they employ the best in their field at the most productive time of their career – then allow them to pursue their passion. This encourages employees to improve their skills and flourish in a challenging and rewarding role. Overall the company has become a talent magnet, without offering excessive remuneration or inducements, relying instead on its underground popularity and cult status. This merit based approach has resulted in a diverse and inclusive workforce, ideally suited to achieve their shared space goals.

"It's really exciting to be a part of something the whole world watches. When we launch a satellite, I can pull up news coverage from countries around the world, and I can say, 'I was a part of that.' Everyone here is working toward the same goal. There is an electricity in the air(5)."
~ *Kyle P. Williams, SpaceX Payload Integration Lead and recent graduate*

SpaceX Headquarters in Hawthorne California, beacon for the young and talented

VERTICAL INTEGRATION

Subsequent to their experience in Russia, SpaceX discovered that most domestic aerospace suppliers maintained similarly punitive profit margins. Basically, purchasing components solely through existing suppliers would undermine or invalidate their goal of reducing launch cost to roughly one tenth of present levels. Worse, when they managed to agree an acceptable price with component suppliers, they would be prone to abruptly rise, removing any possibility of meaningful cost planning. Using SpaceX jargon the supplier had: "gone Russian." Hence to avoid the more predatory vendors they instituted extreme vertical integration (which they call "insourcing"), something almost unheard of in the legacy aerospace industry.

Vertical integration occurs when the parent company purchases one or more of their suppliers, in order to better control their supply chain quality and improve component delivery. However, SpaceX took this one step further by deciding to manufacture as many components inhouse as possible at their fabrication facility in Hawthorne California. This practice has radically reduced component cost and production time, which significantly improved the revenue available to achieve their long term goals. It has been reported that they produce up to 80% of components in-house(6) and still strive to improve this percentage.

"We can produce one [Merlin engine] per day if necessary(7)...We can flex the factory to ship a new booster every 14 days if necessary(8)." ~ *Andy Lambert, SpaceX VP of Production*

9 Merlin Engines, built at SpaceX, ready to be integrated into a Falcon 9 rocket

Vertical integration has certainly improved component delivery and quality, but the main advantage has been cost. Elon Musk stated the complete avionics system for Falcon 9 costs them slightly more than $10 thousand to produce, compared to around $10 million for an equivalent system from conventional aerospace suppliers(9). Similarly SpaceX were quoted $120,000 for an electro-mechanical actuator, which they eventually managed to build inhouse for $3,900.

Last but not least, vertical integration allows the manufacturing process to be amazingly agile. Engineers can amend a component design then manufacture and test it immediately, while they watch. This saves significant time and money, making rapid iteration routine. Given the rate at which SpaceX progress their designs, vertical integration can be viewed as an invaluable resource.

"Rapid iteration is the DNA of the company... a lot of that has to do with [the fact] that we design and manufacture most of the components ourselves. It's very integrated. If the technicians who are putting things together find something that can be improved, they can go straight to the engineer who designed it and that can be fixed on the fly. That's how you can iterate as quickly as we can do today(10)." ~ *David Goldman, SpaceX director of satellite policy*

DECISION MAKING

Good decision making practices are a core resource in the space launch arena where the least delay or misjudgment can lead to the loss of a valuable payload or priceless passengers. Elon Musk sets the tone for SpaceX decision making, which could best be characterized as fast, incisive and coherent.

"He [Elon Musk] makes great decisions with good data. It's irritating that he is right as often as he is(11)." ~ *Gwynne Shotwell, SpaceX COO*

Formal meetings at SpaceX are comparatively rare and usually comprise of no more than five people. In fact, it is quite acceptable for individuals to excuse themself from meetings, if they feel their input isn't entirely relevant.

Rare occasion when Elon relaxed the five per meeting rule, center left President Barack Obama, center right Elon Musk

COO Gwynne Shotwell possesses a more pragmatic management style which complements Musk's more mercurial approach. Together they've steered the company through existential crises and achieved real progress in the space arena, allowing SpaceX to become arguably the most advanced and successful launch company in the world.

"When Elon says something, you have to pause and not immediately blurt out, 'Well, that's impossible,' or, 'There's no way we're going to do that. I don't know how to do that.' So you zip it, and you think about it, and you find ways to get that done...I noticed every time I felt like we were there, we were rolling over [from the steep slope], people were getting comfortable, Elon would throw something out there, and all of a sudden, we're not comfortable and we're climbing that steep slope again. But then once I realized that that's his job, and my job is to get the company close to comfortable so he can push again and put us back on that slope, then I started liking my job a lot more, instead of always being frustrated(12)." ~ *Gwynne Shotwell, SpaceX COO*

Sometimes if they hit a design problem at SpaceX, it's not uncommon for "flash mobs" of engineers to appear in the walkway outside the relevant engineer's cubicle to help brainstorm solutions. Generally, SpaceX engineers have a remarkable degree of autonomy and creative problem solving is encouraged at all levels. Some of the best engineers in the world work or have worked for SpaceX. Occasionally their development projects set out on slightly the wrong track but this always remains a possibility regardless of the care taken during the decision making process. To find the best solution to any problem SpaceX generally attempt to do something first to achieve their goal (simulation, modeling or full scale tests) and learn along the way, rather than wait for panels of inscrutable consultants/accountants to divine their 'perfect' solution.

The road to Mars is undoubtedly a decision-making minefield but the judgment of the crew at SpaceX is extraordinarily well honed – and tempered by their comprehensive experience of building twenty first century spacecraft.

"In the longer term, perhaps in the 10 to 15 year time frame, I'm hopeful that we'll have a craft that can take people to Mars, because the ultimate goal of SpaceX is to develop the technologies that can take humanity to Mars." ~ *Elon Musk, May 2012*

WORK ETHIC

SpaceX employees work hard. They work intensively for long hours, fifty hours a week is expected, although they often do more. As the insider joke goes, "if you skip work on Saturday, don't bother coming in on Sunday!" One employee, Steve Davis (Director of Advanced Projects) has reportedly been putting in 16-hour days every day for years.

"While no one will be forcing you to, you'll end up working crazy long hours, just to keep up with your workload, and because you don't want to leave the place. A phrase I've heard thrown around SpaceX frequently is everyone is their own slave driver. I was frequently there late at night for my job, and I never really felt alone. The factory is always alive and cranking out rockets no matter what time of day or night you go there(13)." ~ *Josh Boehm (former SpaceX intern who graduated to become Head of Software Quality Assurance)*

They say: "a volunteer is worth ten pressed men" and SpaceX essentially consists of 7,000 volunteers... Largely they feel inspired to work harder to achieve something great and worthwhile in the field they love. For instance, SpaceX recruit large numbers of interns who agree to work under these conditions for only basic remuneration. No doubt the work experience they receive at SpaceX is invaluable, but many sign up simply because they want to participate in this drive to democratize space exploration and open the high frontier. According to a recent Pay and Workplace Satisfaction Survey, SpaceX employees rate their company as the best place to work to make the world a better place(14).

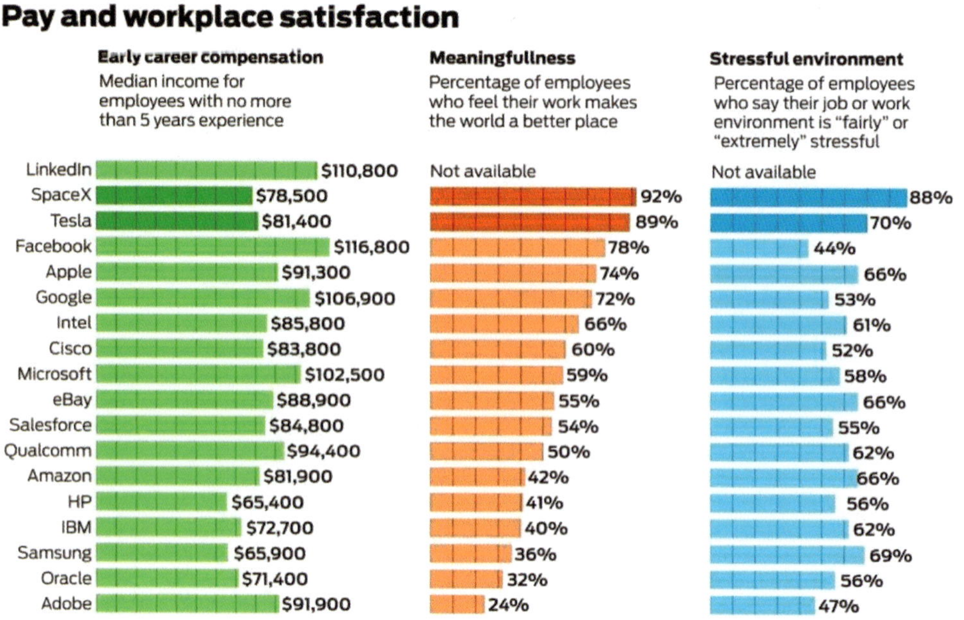

"Many people never achieve their dream job but I can honestly say I have. I enjoy tackling the biggest challenges. Problems do not cause stress to me but rather excite me about the opportunity to get creative and witness the absolute best people have to offer in finding solutions."

"For most people in a manufacturing role, the items produced leave the factory and there is little to no connection with them thereafter or with the customers. At SpaceX, we get to witness the spectacle of launch (and now landing) and see the importance and joy to our customers when we accurately deploy their payloads into orbit. So much hard work goes into achieving all of it, so the personal reward, camaraderie and team spirit for each success is HUGE! I genuinely feel that we're making a contribution to humanity - it's more than just a job(15)!"
~ *Andy Lambert, SpaceX VP of Production*

ADVANCED TECHNOLOGY

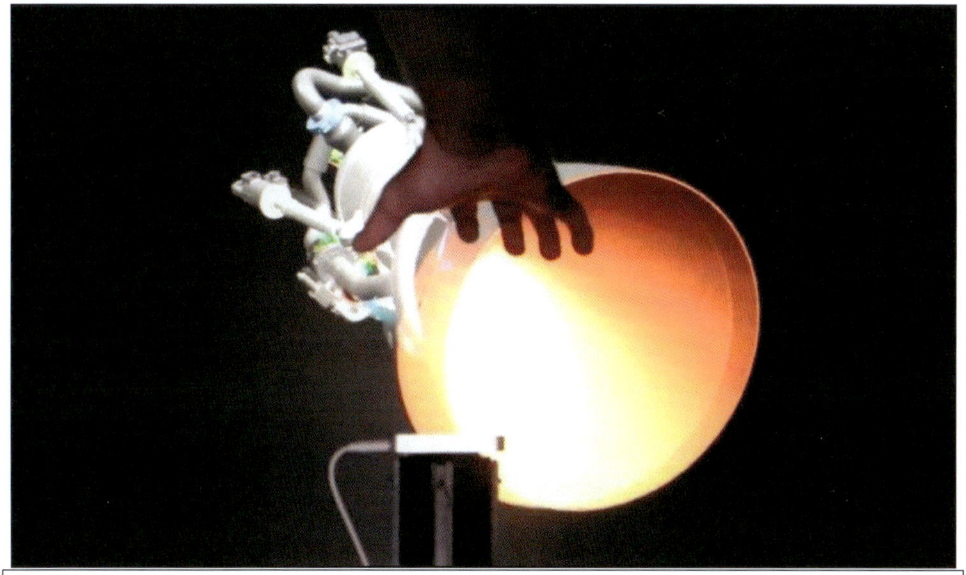

Elon Musk manipulates virtual 3D model of Merlin Engine

Cutting edge technology pervades SpaceX. Motion control CAD(16), laser sintered 3D printing, robotic automation and a lot-lot more that we're not supposed to know about. This approach has given them an incredible advantage over legacy aerospace manufacturers. For example: the airframe of a rocket is usually fabricated by milling and cutting the correct shape and ribs from a solid slab of aluminum. This technique when applied to a large rocket proves horribly complex, time consuming and expensive. However, SpaceX take commercially available strips of aluminum and stir weld them together to form the required airframe. Aluminum can easily catch fire during a conventional welding process but the friction stir welding technique doesn't wholly melt the aluminum, you could almost say it blurs the two parts together.

Because SpaceX build their rockets from scratch every component part and construction technique has been evaluated from a new millennium perspective. Where conventional rockets employ vast and heavy wiring looms for control purposes, they use a few fiber optic cables which are extremely light and can carry thousands of signals simultaneously. Generally everything they do at SpaceX is either highly automated or supported by computers. In fact they spend most of their revenue on R&E (Research and Evaluation) and developing new technologies. You could say SpaceX are an R&D company – that also launch rockets!

"We have never built any two vehicles identically, such is the pace of innovation at SpaceX(17)." ~ *Andy Lambert, SpaceX VP of Production*

SOFTWARE-CENTRIC

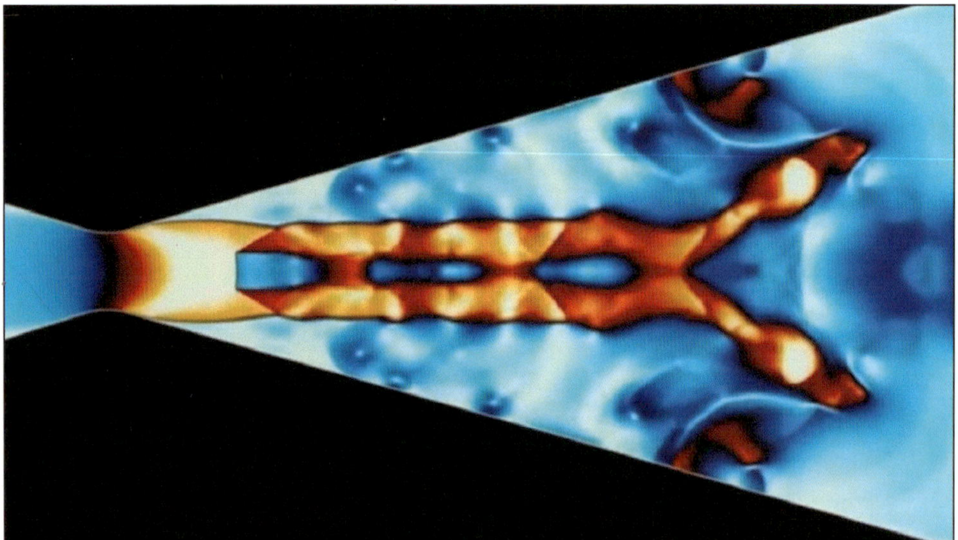

Computational Fluid Dynamics (CFD) simulation of rocket engine combustion, credit SpaceX

Computer software is ubiquitous at SpaceX, its use is as broad and deep as the Pacific Ocean. Elon Musk cut his teeth building Silicon Valley software companies (Zip2 and PayPal) so has a fundamental appreciation of the overwhelming advantages offered by custom built software. Everything from hardware development to production to space operations is facilitated by software developed in-house.

"Something that's worth noting is a lot of what is needed on a rocket or spacecraft is actually software(18)." ~ *Elon Musk*

Linux is used throughout the company, including developer's desktops, which allows them to buy 'out of the box' development tools, like ftrace, gdb, netfilter and iptables, although they have also developed a lot of their own POSIX-based tools(19).

Here's a table to help map some of the software commonly used by SpaceX developers(20): -

Software Overview Table		
Software Tier	**Software Title**	**Software Description**
1	Linux	Basic Operating System (used by office, production and spacecraft)
2	C++	General-purpose programming language
2	LabView	Graphical programming tool (used to visualize vehicle telemetry)
3	Python	High-level programming language
3	GCC	Multi-language compiler
3	Clang	Multi-language compiler
4	Bazel	Automated software development tool

For some companies software is almost an afterthought, whereas at SpaceX software comes first and last.

Artificial Intelligence

"Machines are dumb... ground systems are dumb... We are trying to automate everything... We are doing some AI work to make sure we get humans out of the loop in data analysis to speed up that process(21)." ~ *Gwynne Shotwell*

Yes, it seems SpaceX have a pet AI. Possibly this could be a high machine intelligence software package designed to perform one task extremely proficiently, that's readily available off the shelf. However, SpaceX are able to source technology from two other companies founded by Elon Musk: OpenAI who are developing open source AI software and Tesla Inc. who are creating an advanced AI chip, based on their FSD (Full Self Driving) processor. If this is the case SpaceX's AI technology could soon approach AGI levels (Artificial General Intelligence) which is something they will certainly require for the future (see Chapter 8: How SpaceX Can Source Mars Colony Technology).

One thing's for sure, SpaceX have harnessed some heavyweight computer simulation technology. Before the first flight of Falcon Heavy, Elon was not confident the launch would be a success: "There's a lot that could go wrong, a really tremendous amount. I really like to emphasize that the odds of success are not super high(22).” Fortunately, as it turned out, each phase of the launch was performed successfully – just as their computer simulations predicted.

In the future, if SpaceX decide to string enough advanced FSD processors together, it’s possible they might create the first ASI (Artificial Super Intelligence) – which might possibly regard us as pets!

ROCKET REUSE

Recovered CRS-8 Booster returning to Cape Canaveral on-board ASDS OCISLY (Automated Spaceport Drone Ship - Of Course I Still Love You)

Since the beginning of the sixties space race every stage of the rocket has been discarded in order to lose sufficient weight for the payload to reach its objective. But if these expended stages could somehow be recovered intact and used for subsequent flights, it could dramatically improve our ability to access space. In theory, you would only need to reassemble the rocket stages, refuel and relaunch, virtually eliminating the complex and costly manufacturing process. Normally the propellant (fuel) costs around $0.2m(23) for a Falcon 9 rocket, which is only a tiny fraction of the normal launch cost.

Elon Musk has stated if they manage to fully reuse their rockets this could reduce the launch cost by a factor of 100. For this reason ‘reusability’ has become the mantra at SpaceX. However, landing even the first stage of the Falcon 9 rocket after an orbital mission proved...challenging. Fortunately, SpaceX managed to land a stage at Cape Canaveral in late 2015(2) and now aim to recover and reuse all first stages from Falcon 9 and Falcon Heavy launches. More recently they have managed to refurbish and reuse their Dragon spacecraft (used for cargo deliveries to the ISS) and even recover payload fairings using giant nets suspended above specially adapted support vessels.

SpaceX's specialized fairing catching vessels, Go Ms Chief and Go Ms Tree...

SpaceX regard this ability to reuse flight hardware as vital technology for their journey to Mars. They intend to use a SHL (Super Heavy Lift) rocket to send up to 150 metric tons of cargo to the surface of Mars(25) by reusing every stage, both on the ground and in orbit. SpaceX even plan to return the Mars landing stage to Earth, allowing the hardware to be reused on subsequent Mars missions. Reusability is a revolutionary advance in rocket technology which promises to throw open the doors to space.

"If humanity is to become multi-planetary, the fundamental breakthrough that needs to occur in rocketry is a rapidly and completely reusable rocket...achieving it would be on a par with what the Wright brothers did. It's the fundamental thing that's necessary for humanity to become a spacefaring civilization." ~ *Elon Musk*

Elon Musk
@ID_AA_Carmack Full RUD (rapid unscheduled disassembly) event. Ship is fine minor repairs. Exciting day!

To succeed in something new you can't be afraid to fail. SpaceX engineers (including senior management) actively embrace the risk of failure and actually seem to enjoy it!

"Failure is an option here. If things are not failing, you are not innovating enough(26)."
~ *Elon Musk*

This is similar to the approach used by software developers, who are quite happy for failures to occur during early stage development (Pre-alpha, Alpha and Beta phases). In fact, these failures can be viewed as positive because they signpost any changes that need to be made before they can progress to the next and final iteration.

Likewise, after SpaceX decide the path they wish to pursue on any new project they don't try to analyze everything to the nth degree before they start, which can be counterproductive. Generally, they test some components then assemble them as soon as possible, which allows them to perform an early system test (equivalent to pre-alpha). These early tests produce a treasure trove of information and usually reveal some problems/effects they were previously unaware of. Armed with this new information, they quickly proceed to the next iteration (Alpha phase) which allows more information and iteration through to the final product. This experience-based approach allows them to produce something insightful and practical in the minimum time and cost. But it doesn't end there...

Most engineers accept that new technology won't be perfect first time around, but SpaceX take this one step further. They pursue projects which they feel are unlikely to succeed because they know such failures should better inform their decisions on the best path to proceed. As an early philosopher once said: -

"We learn wisdom from failure much more than from success. We often discover what will do, by finding out what will not do; and probably he who never made a mistake never made a discovery."
~ *Samuel Smiles (Philosopher)*

Originally SpaceX intended to recover their booster stages by parachuting them into the ocean, despite some previous efforts showing this technique was uneconomic (notably NASA's recovery and reuse of the Space Shuttle's Solid Rocket Boosters). However, SpaceX made the attempt and quickly discovered the braking parachutes were being "ripped off" the booster because of the high re-entry speed. Therefore, they needed to find some way to slow the descent before they could deploy the parachutes - like turning the stage around and firing the engines to decelerate (called retropropulsion). The engineers reasoned if they could use retropropulsion, they could possibly dispense with parachutes altogether and attempt to land the Falcon 9 first stage propulsively, using one or more of its existing engines. Elon Musk was cautious whether this idea might work, on the grounds the extra fuel required for landing could reduce the useful payload the rocket carried to orbit. Notably they decided to proceed with tests anyway, ignoring any risk of failure.

What they created was effectively an entirely new rocket, with extra fuel capacity, added performance and improved engine layout. These improvements allowed them to perform first stage recovery with little or no impact on payload capacity, particularly if they landed the stage downrange on an "ASDS" (Autonomous Spaceport Drone Ship).

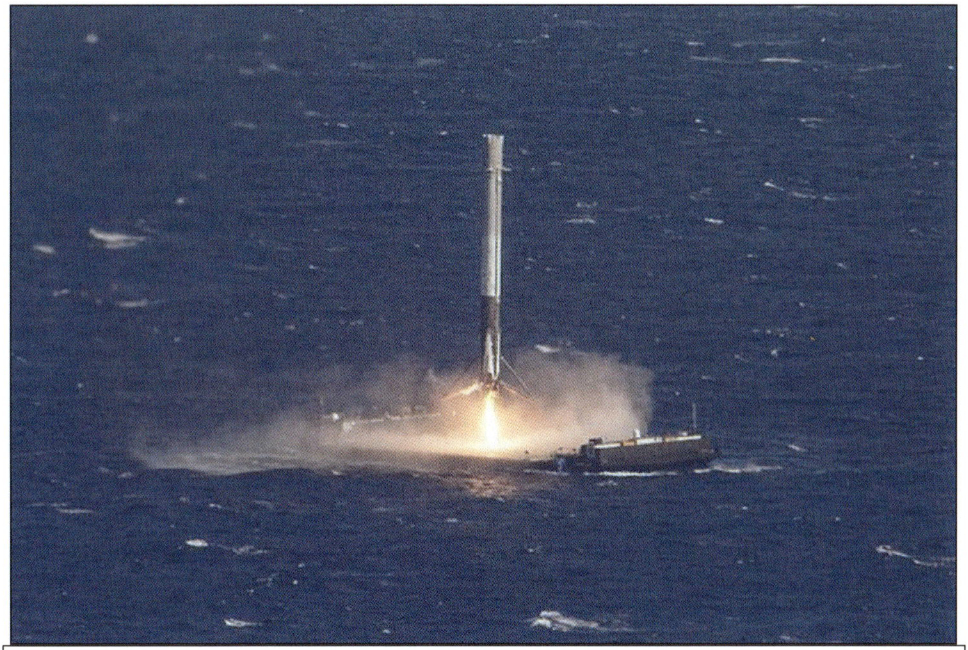

Falcon 9 First Stage landing on ASDS

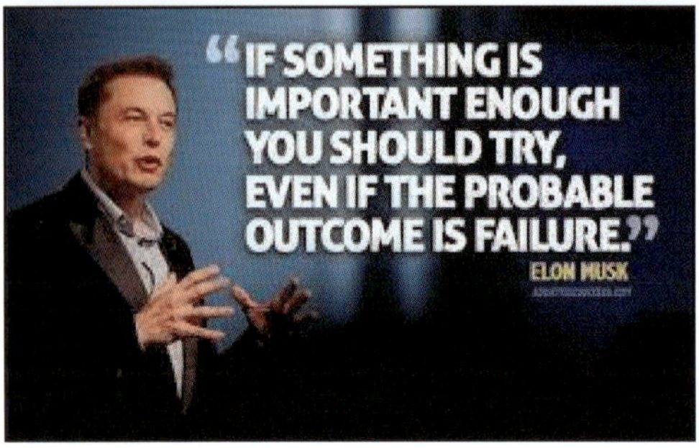
"IF SOMETHING IS IMPORTANT ENOUGH YOU SHOULD TRY, EVEN IF THE PROBABLE OUTCOME IS FAILURE."
ELON MUSK

Ironically this 'damn the torpedoes' approach is probably best illustrated by Elon Musk's philosophy when he decided to start SpaceX. He reasoned the company was more likely to fail than succeed – but decided to go ahead anyway because of the profound benefits offered to humanity by the prospect of colonizing Mars.

Note, once the company was founded, Elon Musk immediately flipped his reservations about the company's viability and became totally committed to its long-term success, no matter the circumstance.

"I don't ever give up. I mean, I'd have to be dead or completely incapacitated." ~ *Elon Musk*

CREATIVE ENGINEERING

Engineers have to be creative by definition but SpaceX engineers take this to a new level. They often embrace development approaches which wouldn't be approved or even considered by more traditional aerospace companies.

Engines and Rockets

When building a rocket from scratch the conventional approach is to develop a number of different sized engines, with each engine capable of producing enough thrust to power each stage of the rocket. This was certainly the approach used by SpaceX on their first rocket, the Falcon 1, which used a single Merlin 1A engine on the first stage and a less powerful but lighter Kestrel engine on the second. However, Falcon 1 lacked sufficient power to launch people (part of their corporate goal) so before it had even flown successfully, SpaceX engineers began to develop a much larger rocket. The design solution they came up with was the Falcon 9, so named because it used nine Merlin 1 engines on the booster stage and a single Merlin 1(fitted with a vacuum exhaust bell) to power the upper stage. Developing a new rocket engine can be time consuming and enormously costly, so by reusing the Falcon 1 main engine they managed to field the Falcon 9 rocket in record time and at a tenth of normal cost, according to NASA estimates(27).

But SpaceX didn't stop there. Most aerospace companies effectively freeze their rocket design after flight tests are complete, to avoid introducing any flaws into an already proven vehicle. However, SpaceX is continually iterating and experimenting with their launch vehicles, while they are being used to launch commercial payloads. The Merlin engine has advanced from versions 1A to 1D improving in performance from 340 kN to 845 KN thrust and now possess the best thrust to weight ratio for any rocket engine in the world. Similarly, Falcon 9 has gone through many major iterations, culminating in Block V. This final design iteration is almost unrecognizable from the first, with the booster stretched to carry more fuel and fitted with landing legs plus grid fins, which enable it to land and be reused 10 times or more without major overhaul. SpaceX even experimented with strapping three Falcon 9 boosters together, creating Falcon Heavy, the most powerful launch vehicle in operation – and 90% reusable. On past performance, this puts them in the driving seat for developing a manned Mars vehicle, which will no doubt require an enormous amount of new and creative engineering.

Starhopper

Starhopper under construction at Boca Chica Texas, credit SpaceX

In late November 2018 something strange happened out at Boca Chica Beach Texas, the site where SpaceX are constructing their own private launch facility. A water tower contractor called Caldwell Tanks(28) had begun to build a 9m wide cylindrical structure made from welded stainless steel, supported on a circular concrete base. Everything pointed to the fact this was indeed a water tower – until it was fitted with landing legs and an aerodynamic nose cone (made from stainless steel sections shipped from Los Angeles). In fact, this was the first prototype Mars vehicle, dubbed the Starhopper, about as far as you could get from a static water tower.

In a bold move SpaceX had contracted out construction of the base section of this test vehicle to building contractors – something unthinkable in the aerospace world. However, Starhopper wasn't designed to carry people, they only needed it to test takeoff and landing; guaranteed to be risky with a completely new engine and airframe, hence disposable. Considering the Starhopper hull was constructed in a little over a month this novel approach likely saved SpaceX a lot of time, difficulty and money, helping them stay on track for their aggressive Mars landing schedule.

"Starhopper flight successful. Water towers *can* fly haha!!" ~ *Elon Musk...*

COURAGE CULTURE

"SpaceX is like Special Forces...we do the missions that others think are impossible. We have goals that are absurdly ambitious by any reasonable standard, but we're going to make them happen. We have the potential here at SpaceX to have an incredible effect on the future of humanity and life itself." ~ *Elon Musk*

For a company as courageous as SpaceX, there are no limits. Top echelon management might set the tone, but everyone has the buzz. They are heroes achieving the impossible each day, despite the odds. This culture of courage has allowed them to go from nothing to the most powerful launch company in the world in less than 16 years. Perhaps the best way to demonstrate their particular brand of courage is through a few examples.

SpaceX technician fixes engine bell prior to launch

In December 2010 SpaceX were preparing to launch the second Falcon 9 mission for NASA, a crucial demonstration flight for their cargo Dragon spacecraft. Mere days before launch they discovered a small crack in the exhaust bell, fitted to the second stage engine. Normal practice dictates the whole bell should be removed and replaced – a costly and time-consuming procedure. However, Elon suggested a more practical course... The crack was relatively minor and located near the outer edge of the spun structure, which made it accessible. The bell was made of niobium alloy, spun to extraordinary fineness, thinner even than a human fingernail along the outer rim. Hence it might be possible to trim the bell, like you would a cracked fingernail, completely removing the flawed section. SpaceX engineers conscientiously ran the numbers and discovered that yes that was a viable solution. The test payload was relatively light (a small wheel of cheese and some cubesats) hence the slight reduction in engine efficiency shouldn't affect the overall performance of the mission. So a SpaceX technician, who happened to be afraid of flying, flew to the Cape post haste. Using a pair of tin snips, he managed to trim out the split section, restoring the vehicle to flightworthy condition, only a day before launch. And the rest is history; NASA was both heartened and impressed by the test mission, which proved a complete success, with minimum delay(29).

Here's another example: while preparing for a Falcon 9 launch, an error was spotted in a software file, just hours before liftoff(9). So SpaceX engineers immediately fixed the file then confirmed it worked correctly, using a Falcon 9 test bed at their Hawthorne Headquarters. Happy the fix worked, they promptly uploaded the software file to the rocket, all within a half hour of when the fault was originally discovered. If something similar had occurred prior to a Space Shuttle launch, the flight would have stood down for at least three weeks, while the problem was examined from every possible angle with all of the contractors concerned.

Amos 6 explosion during Falcon 9 test at SLC-40, Cape Canaveral

Of course, things don't always go as planned when you are bold. For instance, in September 2016 SpaceX lost a Falcon 9 launch vehicle during a routine test firing, which also destroyed the Israeli communication satellite Amos 6. After performing an exhaustive investigation(30) SpaceX determined a change to the helium loading procedure (helium is used to pressurize the propellant tanks) caused the propellant to ignite, resulting in a loss of vehicle. The procedure had been changed to allow faster loading of pressurant and propellant, in order to minimize temperature rise (and expansion) of the deep cryo fuel. However, this crash loading procedure had inadvertently caused the helium containers, called Composite Overwrapped Pressure Vessels (COPVs), to buckle... Because the highly pressurized helium COPVs were suspended in the liquid oxygen propellant tank, this failure caused the carbon composite to react with the oxygen, resulting in a rather spectacular explosion. As we've already discovered, failures like this actually teach a great deal and if nothing else defined the limits of Falcon 9 operation and a possible weakness. Currently SpaceX have transitioned to "COPV 2.0" – possibly the most tested and safest pressure vessel in the world. In addition, their next generation launch vehicle, called Starship, will dispense entirely with propellant pressurizing COPV's, effectively removing this component as a source of failure.

Arguably the best way to demonstrate courage is through sacrifice.

"Right now, in my department we're forming up a small group that will be advocates for future astronauts. And that group could grow into an office where the astronauts are managed inside of SpaceX(31)." ~ *Ken Bowersox, former Vice President of Astronaut Safety and Mission assurance at SpaceX*

Seems working for SpaceX requires another level of commitment... SpaceXers certainly have courage – which will be a key resource for Mars.

SPACE COMMUNITY

SpaceX team celebrate completing another Dragon Spacecraft, credit SpaceX

For any viable community to exist in space, everyone must be highly interdependent, cohesive and productive. Their goals must align, with each person willing to support their peers 100% for the good of all. This ethos is already practiced at SpaceX where individuals are motivated and inspired to pursue purely space related goals, in concert with everyone else who works in this nascent space community. SpaceX personnel work long hours in challenging roles, but this generates strong camaraderie, which is continually reinforced as they achieve their shared goals. Notably these successes seem to occur with increasing regularity, as their numbers grow.

"The job satisfaction and team camaraderie is like nowhere else. Every time there is a launch, everyone crowds around mission control and cheers it on. Getting your mission patch after a launch was always a very satisfying feeling. If there was ever a failure, you definitely felt it in the air, but it wouldn't stop any of us from working or demotivate us. The pride for the company is unbelievable. How do you spot a SpaceX employee? They're covered head to toe in company swag, and keeps mentioning how they work for SpaceX(13)." ~ *Josh Boehm, former Head of Software Quality Assurance at SpaceX*

CONCLUSION

1. Mars is so challenging that SpaceX have been forced to take a holistic approach i.e. fine tune everything to achieve their goal. This has significantly assisted the space effort and made them highly successful as a commercial launch provider.

2. Pushing the boundaries of what's possible is inherently risky and not without cost. Fortunately, SpaceX remains a private concern, hence better able to absorb such setbacks compared to a public company or space agency.

3. SpaceX are no longer an aerospace company, their advanced philosophy and futuristic approach have allowed them to evolve into a fully functional "space company," ready for the challenges of Mars. Likely they will continue to accumulate significant advances and redefine our attitudes on how we approach space exploration, as they ascend to reach their goal.

Succinctly put: Mars is not a case of 'if' as 'when' for SpaceX.

[1]https://www.teslamotors.com/en_GB/about
[2]https://www.cnet.com/news/elon-musk-gas-should-cost-10-per-gallon/
[3]https://twitter.com/SciGuySpace/status/1166477457639317504
[4]https://www.teslarati.com/tesla-spacex-best-employers-2019-elon-musk/
[5]https://news.erau.edu/headlines/whats-it-like-to-work-for-spacex-alumnus-kyle-williams-tells-all
[6]https://youtu.be/yBPV73Fq820?t=1897
[7]https://www.reddit.com/r/spacex/comments/8el28f/i_am_andy_lambert_spacexs_vp_of_production_ask_me/dxw8ix3/?context=3
[8]https://www.reddit.com/r/spacex/comments/8el28f/i_am_andy_lambert_spacexs_vp_of_production_ask_me/dxw83bx/?context=3
[9]ISBN: 9780062301239, "Elon Musk: Tesla, SpaceX, and the Quest for a Fantastic Future" by Ashlee Vance
[10]https://spacenews.com/three-rules-for-building-a-megaconstellation/

[11]https://www.bloomberg.com/news/features/2018-07-26/she-launches-spaceships-sells-rockets-and-deals-with-elon-musk
[12]https://www.youtube.com/watch?v=Dar8P3r7GYA&feature=youtu.be&t=514
[13]https://www.quora.com/What-is-it-like-to-work-at-SpaceX
[14]http://www.sfgate.com/business/article/For-Millenials-in-tech-work-is-about-both-6864194.php
[15]https://www.reddit.com/r/spacex/comments/8el28f/i_am_andy_lambert_spacexs_vp_of_production_ask_me/dxw5hoy/?context=3
[16]https://www.linkedin.com/pulse/virtually-designed-products-3d-printers-21st-century-schilling?forceNoSplash=true
[17]https://www.reddit.com/r/spacex/comments/8el28f/i_am_andy_lambert_spacexs_vp_of_production_ask_me/dxw77tf/?context=3
[18]https://youtu.be/uBaLYDbk4fY?t=1105
[19]http://lwn.net/Articles/540368/
[20]https://www.youtube.com/watch?v=t_3bckhV_Yl&feature=youtu.be
[21]https://youtu.be/X1mp1j0ef8c?t=1587
[22]https://arstechnica.com/science/2018/02/at-the-pad-elon-musk-sizes-up-the-falcon-heavys-chance-of-success/
[23]http://www.foxnews.com/science/2015/12/22/spacex-ceo-elon-musk-eyes-city-on-mars-after-successful-rocket-landing.html
[24]https://www.youtube.com/watch?v=1B6oiLNyKKI
[25]https://en.wikipedia.org/wiki/Starship_(rocket)
[26]https://www.logomaker.com/blog/2013/05/13/14-inspirational-quotes-for-small-business-from-elon-musk/
[27]https://arstechnica.com/science/2017/07/spacex-urges-lawmakers-to-commercialize-deep-space-exploration/
[28]http://www.caldwellwatertanks.com/
[29]https://naturallyfundamental.com/spacex-tin-snips-rocket-fix/
[30]https://www.spacex.com/news/2016/09/01/anomaly-updates
[31]https://www.youtube.com/watch?v=rSSeC5ka6uA&feature=youtu.be&t=

Chapter 5: How SpaceX Can Pay for Mars

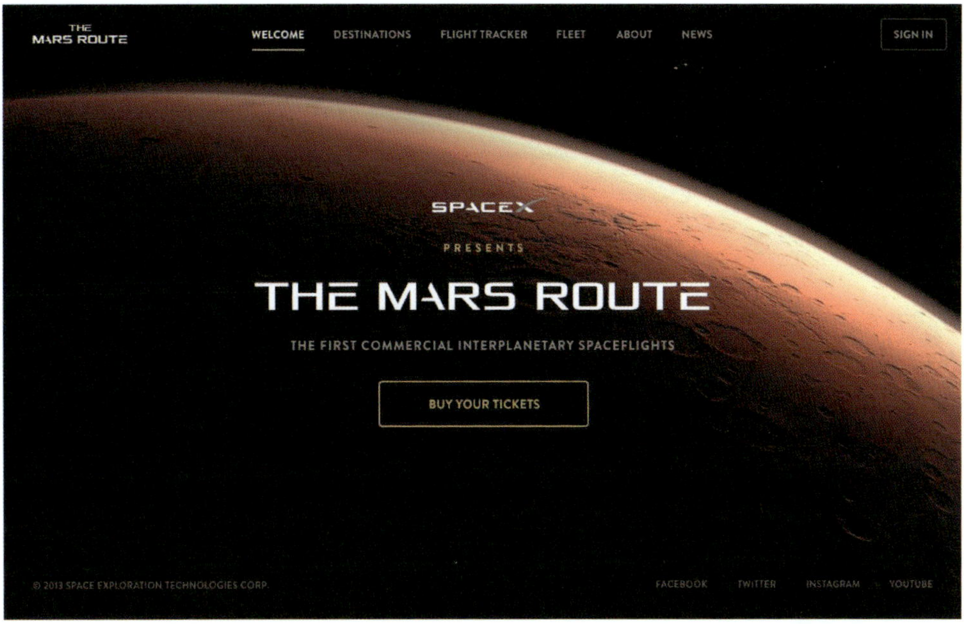

SpaceX want to take us to other worlds and have their sights squarely set on Mars. However, they will need a mighty war chest to mount a successful campaign on Mars. Developing the necessary technology to transport large quantities of colonists and cargo could require tens of billions – or at least a steady revenue stream of billions.

SpaceX managed 21 launches in 2018, achieving a far greater launch rate than any other company in the world, so they are certainly receiving significant revenue. However, the world launch market appears fairly static, arguably capped at ~$5bn p.a., which means they could find it quite challenging to squeeze the billions they need from this seemingly inelastic market. Given what we know of SpaceX, it should come as no surprise they intend something truly remarkable, if not revolutionary for a space company; they intend to finance this project themselves, through purely commercial means.

THE SPACEX FINANCE PLAN

Make Space Pay

At IAC 2017 Elon Musk gave us some insight into how they plan to finance the next generation rocket required for Mars colonization, which they call "Starship." After studying projected earnings and outgoings, he confirmed it should be possible to finance its development through purely commercial means, i.e. using revenue from their commercial launch business. However, they will need to exact certain extreme measures to ensure this finite revenue stream proves sufficient to allow them to transition into a multiplanetary transport company.

Cannibalize Own Products

In 2019 SpaceX plan to freeze development of the Dragon 2 and Falcon rockets, allowing their development teams to focus fully on Starship. Then after a large stable of Dragon 2 and Falcon Block 5 vehicles have been produced, they intend to cease production of these workhorse vehicles entirely. This should free the production crew to work exclusively on Starship, which they intend to produce in volume. Engineering a staged withdrawal from existing vehicles should allow SpaceX to switch all commercial revenue to this make or break Mars project.

If all goes well their fleet of Falcons and Dragons should be capable of fulfilling all launch commitments for a number of years (possible with rapid reuse capability) until their new rocket comes online. Then this exceptionally large vehicle will shoulder all launch services, including Earth orbit, cislunar and Mars colonization missions.

"SpaceX will prob build 30 to 40 rocket cores for ~300 missions over 5 years. Then [Starship] takes over & Falcon retires(1)." ~ *Elon Musk*

Reality Check

At first glance, it seems incredibly that a lone aerospace company could possibly build their own Mars rocket without some kind of government support – but with sufficient will and money you can do anything. Judging by results, SpaceX certainly have the resolve to achieve great things, so let's delve a little deeper into their finances to see what revenue streams they have available.

A NEW KIND OF LAUNCH SERVICE

In the past, most aerospace companies have taken an extremely expensive approach to providing launch services, stemming from the low number of providers and general unwillingness to find low cost alternatives to the status quo. However, SpaceX have implemented new methods to build and operate launch vehicles then chose to pass on these cost improvements to customers in order to stimulate the space economy. This radical reduction in launch costs has resulted in substantial improvement to their market share in all spheres: civil, commercial and military.

"So we do expect to see a steady reduction in prices, and we already have reduced prices from where they were, from about $60 million to about $50 million for a re-flown booster(2)." ~ *Elon Musk*

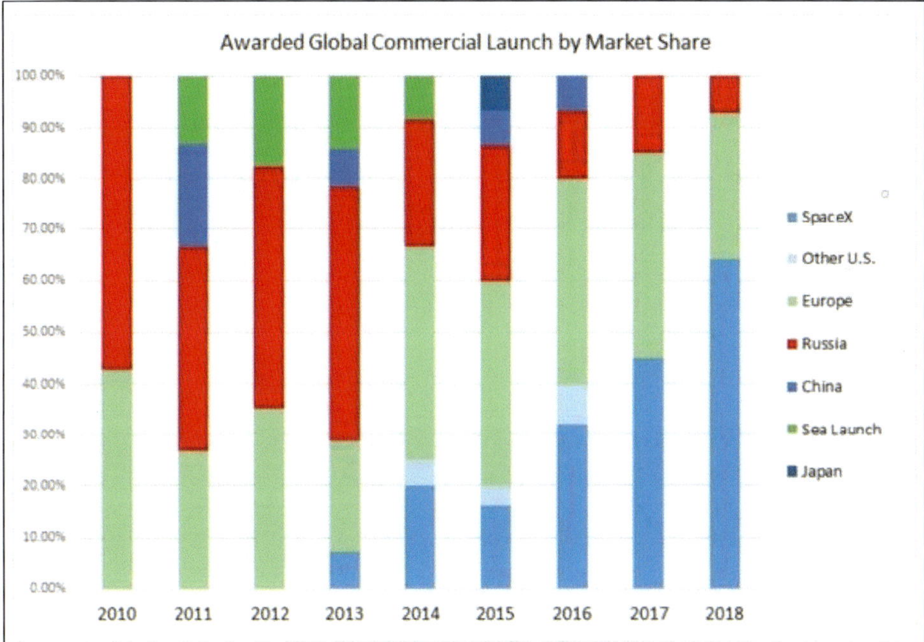

Figure 1: Global Commercial Market Share

SpaceX used Falcon 9 to achieve commercial launch supremacy in 2018

Legacy launch providers are finding it increasingly difficult to compete due to the price disparity caused by this new and disruptive approach. SpaceX also have the capability to launch every two weeks or less, using their reusable boosters, which means they can outperform the competition on both price and delivery.

"...we may be able to get down to a marginal cost for a Falcon 9 launch under five or six million dollars(2)." ~ *Elon Musk*

In the medium term it seems likely SpaceX will absorb most to the commercial market, perhaps even exceeding the existing $5bn market cap, if their stimulus to the space economy results in increased demand. Certainly more commercial satellites are being launched year on year, largely due to expanded interest in the commercial potential of space and industry wide reduction in launch costs – both arguably attributable to the success of SpaceX.

It might be helpful to explore all the possible areas of business that are opening up to SpaceX, to appreciate what is financially possible.

SATELLITE LAUNCH SERVICES

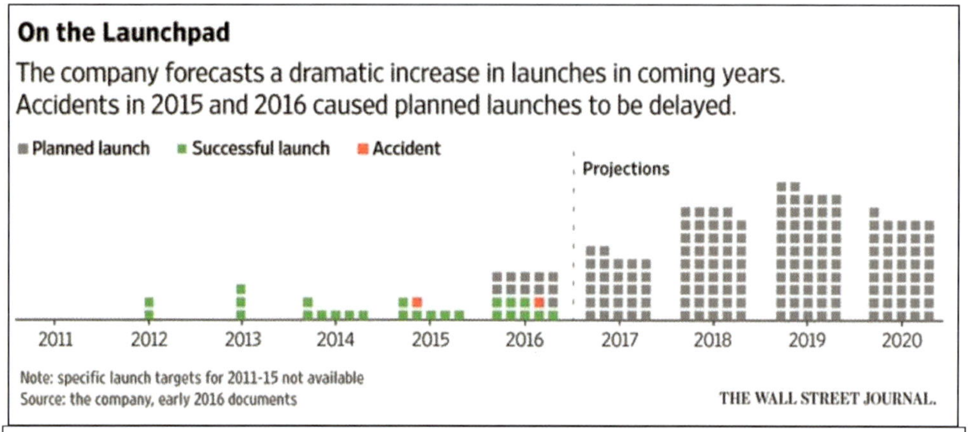

SpaceX launch targets, from early 2016

At present Falcon 9 has a payload capacity of 22.8 mt to LEO and 8.3 mt to GTO which means it can comfortably handle most satellite payloads. SpaceX currently plan to perform 11 satellite launches (commercial, civil and defense) plus another 5 dedicated Starlink launches (see Starlink Constellation below) on Falcon 9 during 2019.

Exceptionally heavy satellite payloads will be handled by Falcon Heavy which has a maximum payload capacity of 63.8 mt to LEO and 26.7 mt to GTO (when flown in expendable mode). Realistically SpaceX should perform only 2 such Falcon Heavy launches in 2019, due to reduced demand for these unusually heavy payloads.

COMMERCIAL AND DEFENSE

For commercial launch services, Elon suggests they charge full price ($62m) for a new Falcon 9 and ~$50m(2) if using a flight proven booster. At present Falcon 9 boosters average 3 flights, which suggests the average price to launch commercial satellites should be ~ $54m.

2019 Satellite Launch Revenue			
Launch Type	**Launches**	**Launch Cost**	**Revenue**
Falcon 9 (commercial)	6	$54m	$324m
Falcon 9 (defense)	3	$96m(3)	$288m
Falcon Heavy (commercial)	1	$90m(4)	$90m
Falcon Heavy (defense)	1	$165m(5)	$165m
Total Revenue (est.)			**$867m p.a.**

In the medium term, SpaceX intend to: "hit a launch cadence of one or two a month from every launch site we have(6)." They also plan to operate two launch sites at Cape Canaveral (LC39-A and SLC-40) and one at Vandenberg Air Force Base (SLC-4E). However, SpaceX could struggle to find customers for more than forty commercial satellite launches per year (roughly comprised of 20 LEO and 20 GTO flights) in the medium term. Hence it seems likely most of this excess capacity will be used to launch SpaceX's own internet satellites (see 'Starlink Constellation' below) and an increasing number of defense payloads. For commercial flights, the Block 5 booster can be reused between 10 to 100 times, which implies the flight proven cost, of ~$50m, will become the norm (first and second flights will probably be reserved for civil/military missions at a higher cost).

In the medium term Falcon Heavy could launch one commercial satellite and one defense payload per year on average, however, these defense missions promise to provide increased revenue. Government agencies usually require extra services (such as mission assurance, production transparency, vertical payload integration etc) which are usually reflected in the price(3).

Medium Term Satellite Launch Revenue			
Launch Type	**Launches p.a.**	**Launch Cost**	**Revenue**
Falcon 9 (commercial)	36	$50m	$1,800m
Falcon 9 (defense)	3	$96m(3)	$288m
Falcon Heavy (commercial)	1	$90m(4)	$90m
Falcon Heavy (defense)	1	$165m(5)	$165m
Total Revenue (est.)			**$2,343m p.a.**

In 2021 SpaceX should begin to use Starship to launch telecom payloads. Likely this extraordinarily powerful rocket would be used to replace Falcon Heavy on commercial flights. Typically, it should command premium revenue, similar to Falcon Heavy, except with lower operating cost due to full reusability.

COMMERCIAL RESUPPLY SERVICES (CRS)

Dragon berths with the ISS

SpaceX currently operate a return cargo service to the International Space Station (ISS) using their Dragon spacecraft launched on Falcon 9. Three such resupply flights are planned for 2019, under an extension to the original Commercial Resupply Services (CRS) contract agreed with NASA. Gwynne Shotwell gave an estimated price of $150m(7) for these flights, which is higher than the Falcon 9 list price but includes the cost for a Dragon spacecraft plus mission assurance overheads for both rocket and spacecraft. NASA require even greater levels of scrutiny for Dragon as it does for Falcon 9, to ensure there are no mishaps with the ISS, which took 10 years and $150bn to build and usually carries a crew complement of six multinational astronauts.

2019 CRS Launch Revenue			
Launch Type	**Launches p.a.**	**Launch Cost**	**Revenue**
Falcon 9	3	$150m	$450m
Total Revenue (est.)			**$450m p.a.**

SpaceX plans to complete these extension flights in early 2020, generating relatively high income. Note: the income from these missions should increase with the introduction of Falcon 9 Block 5 (which is reuse optimized) and if they manage to land Dragon on a large floating pad, which promises to further reduce refurbishment costs.

CCP AND CRS-2

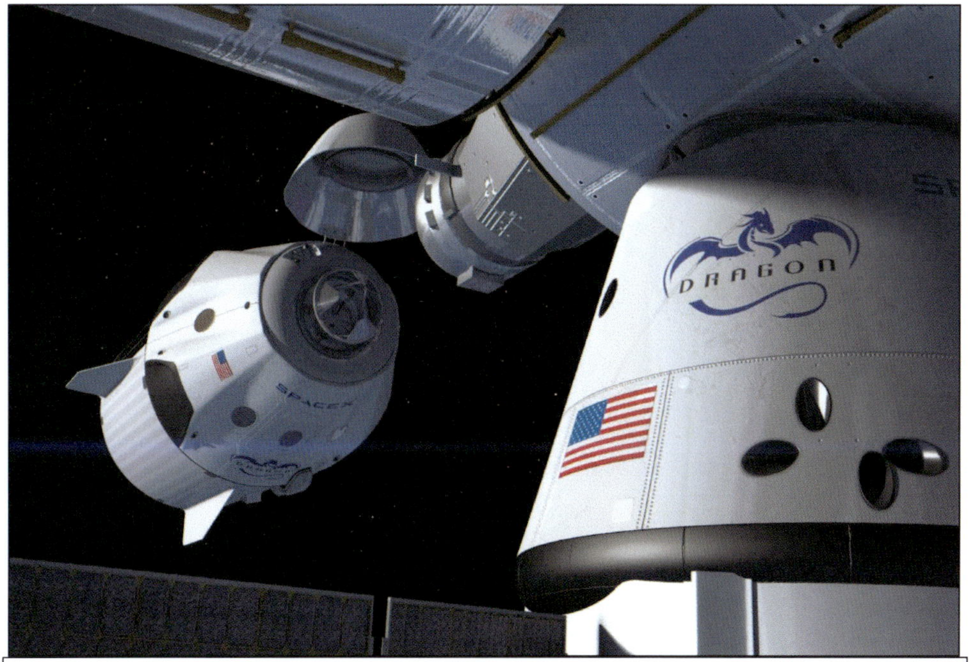

Dragon 2 Docks with the ISS, credit SpaceX

In 2020 SpaceX will begin to transport cargo and crew to the ISS under their Commercial Crew Program (CCP) and Commercial Resupply Service 2 (CRS-2) contracts with NASA. These missions are quite similar because they both use a Falcon 9 Block 5 to send a Dragon 2 spacecraft to the ISS.

NASA's Office of Inspector General reported: "when compared to the cost of each contractor's final CRS-1 mission, SpaceX's average pricing per kilogram will increase approximately 50 percent," while confirming "SpaceX's total upmass capability from Dragon 1 to Dragon 2 did not change." In other words CRS-2 flights cost around 50% more than the CRS extension flights ($150m), or ~$225m.

Reportedly this Dragon 2 price increase is due to conversion costs for hauling cargo (spacecraft was originally designed to carry people), increased pressurized volume, longer duration missions and improved cargo handling capabilities(8). However, it seems likely SpaceX underestimated the cost to develop Dragon 2 during the original bidding process and have now chosen to recoup these losses by increasing revenue from operational missions. In 2014 SpaceX was awarded $2.6bn to develop their crew launch system (including demo flights) while Boeing was awarded $4.2bn to develop a similar capability (excluding demo flights) for NASA's Commercial Crew Program(9). It seems SpaceX significantly underestimated the additional expense from all the safety reviews, testing and certification processes required by NASA for crew missions(10), whereas Boeing, with their greater experience dealing with the administration, provided a more practical bid. For once SpaceX's drive to reduce the cost of space access rebounded but fortunately they should soon commence Dragon 2 missions, to make up any shortfall.

"We've spent actually, I think, quite a lot more than expected [on Crew Dragon]– probably on the order of hundreds of millions of dollars more(11)." ~ *Elon Musk*

To produce a conservative estimate for expected revenue from these missions, let's assume crew flights are similarly priced to cargo, because they use the same launch vehicle and spacecraft (i.e. Falcon 9 and Dragon 2). Hence cargo and crew flights to the ISS should produce the following revenue in the medium term: -

Medium Term CCP & CRS Launch Revenue			
Launch Type	**Launches p.a.**	**Launch Cost**	**Revenue**
Falcon 9 (cargo)	3	$225m	$675m
Falcon 9 (crew)	2	$225m	$450m
Total Revenue (est.)			**$1,125m p.a.**

Interestingly, NASA intend to increase the number of crew onboard the ISS to seven, which implies they could require more flights per year for both cargo and crew.

COMMERCIAL PASSENGER & CARGO TRANSPORT

In the medium term, Bigelow space stations at LEO(12) should require passenger transport flights, which would provide SpaceX an additional stream of revenue. A return ticket to LEO could cost slightly less than $10m for high flight rates(13) and Dragon seats up to seven people, which suggests each flight should cost around $70m. Robert Bigelow (Bigelow Aerospace CEO) stated he will require 24 flights per year to LEO(14), to transport both passengers and goods. If we assume cargo flights have comparable cost to passenger's (both use similar hardware) this suggests the following revenue could be realized from commercial orbital services in the medium term: -

Medium Term Commercial Passenger and Cargo Revenue			
Launch Type	**Launches p.a.**	**Launch Cost**	**Revenue**
Falcon 9 (cargo/passengers)	24	$70m	$1,680m
Total Revenue (est.)			**$1,680m p.a.**

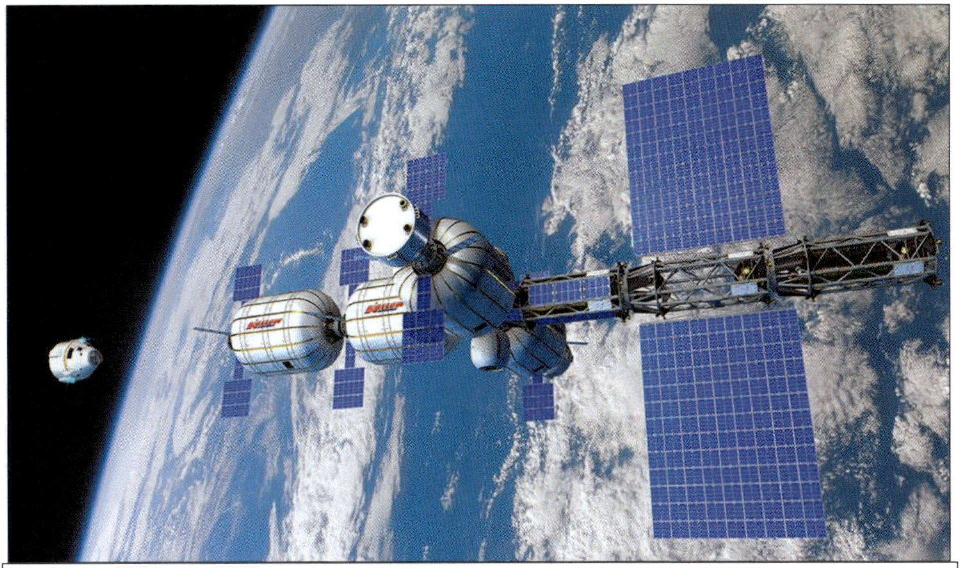

Bigelow Aerospace Alpha Station with Dragon 2 on approach

Currently only two other companies appear capable of mounting such flights: Boeing and Blue Origin. However, Boeing's CST-100 Starliner uses a disposable Atlas V rocket which makes it uncompetitive and Blue Origin seem unlikely to develop an orbital passenger vehicle in less than five years, which effectively gives SpaceX free rein in this arena. After SpaceX won a hard fought contest with Blue Origin for the right to use the historic Apollo launch pad at the Cape, Elon Musk famously observed: -

"If they [Blue Origin] do somehow show up in the next 5 years with a vehicle qualified to NASA's human rating standards that can dock with the Space Station, which is what Pad 39A is meant to do, we will gladly accommodate their needs. Frankly, I think we are more likely to discover unicorns dancing in the flame duct."

In October 2019, NASA announced they will fund commercial destinations in LEO, through their NextSTEP Program(15). Likely Bigelow will be one among many to bid for this commercial space station work, but whoever wins will likely use SpaceX, due to their lower launch prices. Assuming these commercial stations are a success, we could reasonably expect all ISS operations to transfer to them in the next few years, with NASA becoming an anchor tenant.

CISLUNAR LAUNCH SERVICE

Our moon is becoming an increasingly attractive destination for a variety of customers and SpaceX seem ideally placed to tap into this potential market. At present there appears to be three leading prospects for cislunar business, all promising substantial revenue.

Circumlunar Tourism

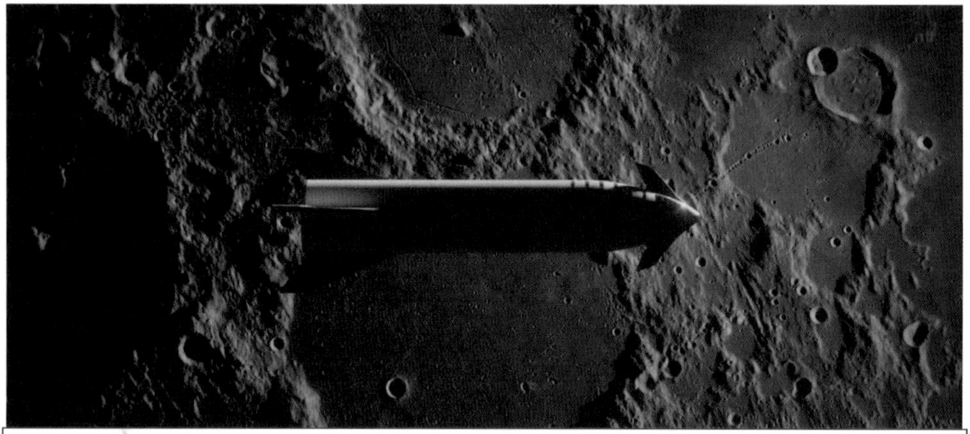

Starship spacecraft cruises past moon, credit SpaceX

In 2017 SpaceX announced they would perform cislunar flights for more adventurous space tourists, commencing 2023. They intend to send their new Starship spacecraft (see Chapter 6: How SpaceX Can Travel To Mars) on a free return trajectory around the moon, to ensure its safe return to Earth. This spacecraft should provide passengers with all the comforts of a space liner, at a fraction of current launch costs, due to full and rapid reusability.

SpaceX aim to launch at least one such mission per year, using flight proven hardware. It has been reported the price for these flights should be ~5% of the total development cost, or approximately $250m per flight(16). Note: SpaceX plan to perform these missions in the mid-2020s but have already generated significant revenue from booking fees, with the balance likely paid before launch.

Cislunar Space Station

In the medium term (2023-24), NASA plan to establish a cislunar habitat called the Lunar Orbital Platform-Gateway (LOP-G), formerly known as the Deep Space Gateway (DSG), which they intend to use as the base of operations for a Deep Space Transport (DST) spacecraft(17). NASA will require commercial transport services to support these operations, similar to the commercial services supplied to the ISS(18). No doubt NASA will opt to use Dragon 2 and Falcon Heavy for these flights because they prefer flight proven hardware and Starship is unlikely to be certified by NASA at this time.

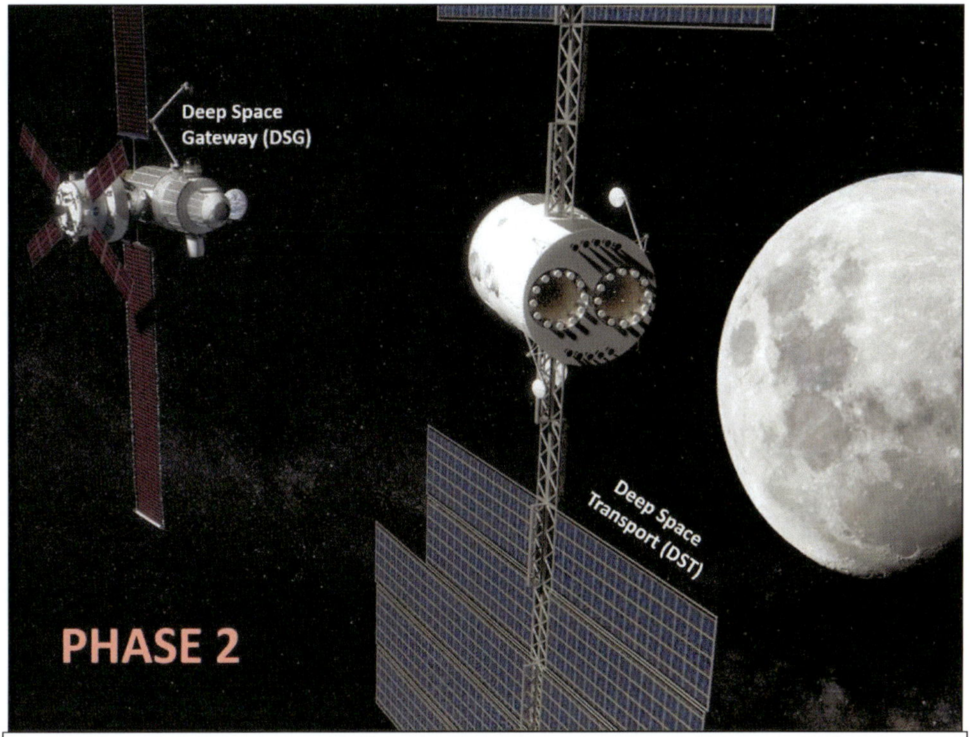

Deep Space Transport spacecraft approaches Deep Space Gateway/LOP-G, credit NASA

By comparison, the projected price for NASA's Space Launch System (SLS) is $1bn to launch cargo(19) or $2bn(20) for crew, so SpaceX should be highly competitive pitching at ~$325m ($100m more than existing LEO missions to cover increased operating cost for Falcon Heavy and long duration missions).

Lunar Lander

SpaceX are studying whether existing Dragon technology can be developed into a lunar landing system for NASA's Artemis Program. This system would consist of a lunar lander (comprised of a descent and ascent stage), a reusable tug to transfer the lander between LOP-G and low lunar orbit, and a refueling tanker, all launched on separate Falcon Heavy vehicles. Given the similarity to LOP-G operations, SpaceX could offer lunar lander, tug and tanker flights for $325m each (basic LOP-G price) assuming they can reuse the Falcon Heavy side boosters. Somewhat higher priced than equivalent ISS missions but likely to be highly competitive with any comparable system based on SLS/Orion.

Artemis lunar lander, credit NASA

Cislunar Revenue

Assuming two LOP-G, lunar lander and tourist flights are performed each year the revenue for cislunar flights should be: -

Medium Term Cislunar Launch Revenue			
Launch Type	**Launches p.a.**	**Launch Cost**	**Revenue**
Starship	2	$250m	$500m
Falcon Heavy (LOP-G Passenger and Cargo)	2	$325m	$650m
Falcon Heavy (LOP-G tug/tanker vehicle)	2	$325m	$650m
Falcon Heavy (Lunar lander)	2	$325m	$650m
Total Revenue (est.)			**$2,450m p.a.**

SpaceX could face stiff competition for these flights from Blue Origin, who intend to operate their own cislunar launch vehicle (called New Glenn) in the late 2020's. However, by then SpaceX should have established their reputation as a reliable service provider, plus their ultra-low operating costs could allow them enormous flexibility on price.

FINANCING MARS TRANSPORT

Short Term Launch Revenue

Given all the above estimates for earnings, here's what we can reasonably project for SpaceX revenue in the short term: -

2019 launch Revenue	
Satellite	$867m
CRS	$450m
Total Revenue (est.)	**$1,317m**

This suggests SpaceX is generating sufficient revenue to support development of their Starship vehicle, assuming a reasonable profit margin. After the SES 10 satellite was launched, using the first flight proven booster, Elon Musk stated it cost them at least $1bn to develop rocket reusability(21). Likewise, after the Falcon Heavy demonstration flight, Elon estimated they had spent at least $500m(22) developing this heavy lift capability. Both these efforts were financed solely by SpaceX, so they generated these funds through normal commercial operations, following a total of 48 Falcon 9 flights.

However, SpaceX have pursued many other projects in parallel over the same launch period. They also renovated 3 launch centers, began work on a new development facility at Boca Chica Texas and considerably expanded operations at their Headquarters in Hawthorne California, main test site in McGregor Texas and Cape Canaveral.

"We keep extending the facility in the Hawthorne area, in Los Angeles County. I feel like I sign a lease every month or six weeks for a new facility. We should be able to produce 40 cores a year in that factory(6)." ~ *Gwynne Shotwell*

When the first Falcon 9 and Dragon were built back in 2009, SpaceX employed 620 people(23). Fast forward to 2017 and they managed to perform 18 successful flights (including 4 Dragon resupplies to the ISS) and employed over 6,000. Basically SpaceX funded this massive expansion with the revenue from 48 launches over 8 years - and plan to perform another 14 launches in 2019 alone (plus additional Starlink flights).

Note: any returns derived from launch services can be regarded as a surplus because SpaceX receive ~$0.5bn per year from NASA for development work carried out under the Commercial Crew Program. Essentially CCP money is used to keep the lights on at SpaceX, allowing any addition commercial profit to be used as disposable income.

Considering SpaceX's profitability, it's shocking to consider their launch prices are among the lowest in the world. Instead of exploiting their extraordinary cost savings they have decided to pass on more than their fair share to customers. This has allowed SpaceX to generate significant returns, while succeeding in their overall goal to reduce the cost of space access.

"...we're trying to make space accessible to everyone. We want it to be such that if you want to go to orbit or beyond, then you can do so. We want to open up space for humanity, and in order to do that, space must be affordable(24)." ~ *Elon Musk*

Medium Term Launch Revenue

No doubt the cost to develop a Mars capable vehicle will increase over time, so let's examine SpaceX revenue over the next few years: -

Medium Term launch Revenue	
Satellite	$2,343m
CCP & CRS	$1,125m
Commercial Passenger and Cargo	$1,680m
Cislunar	$2,450m
Total Revenue (est.)	**$7,598m p.a.**

Assuming stable market conditions, SpaceX could have enormous financial leverage in the medium term. If some revenue streams fail to materialize, they should still have sufficient return to independently develop a Mars launch vehicle and necessary infrastructure.

Figure 2: NASA's Awards for Commercial Cargo and Crew Activities Through 2024

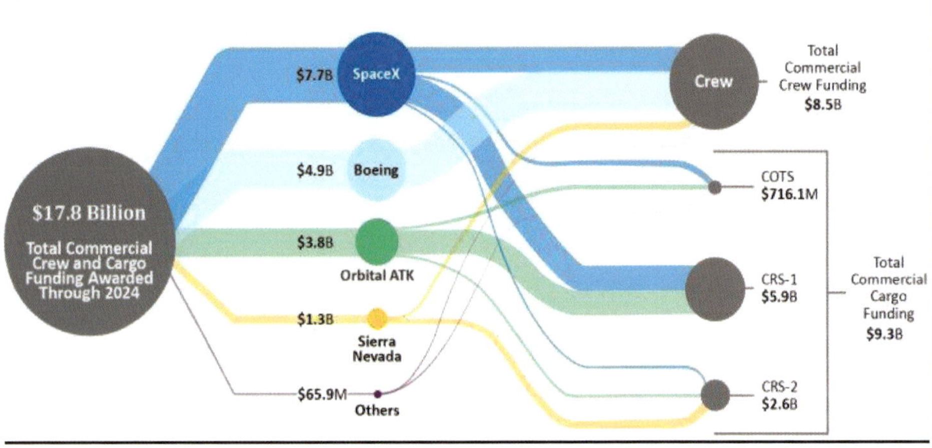

Source: NASA OIG analysis of Agency commercial activities and awards.

Realistically SpaceX revenue and return could double for the next few years, assuming profit margins are maintained. The implied returns from this unparalleled space activity might appear high, until you consider how this money is actually used. Any operating surplus is reinvested in the company to renovate, expand or create new infrastructure and accelerate space development. This creates countless well paid jobs(25) in a variety of roles; such as construction, test, research, fabrication etc. In the long term this pivotal investment should help fashion a space economy of unprecedented scale, no doubt improving the way we live on Earth. Projected profits are substantial, however, SpaceX have gone to extraordinary lengths and worked incredibly hard to earn them. Basically they are warranted because if you want to achieve anything meaningful in the space arena, prices begin in billions.

To conclude, the succinct answer to the question: "how can SpaceX get to Mars" is "one launch at a time."

STARLINK CONSTELLATION

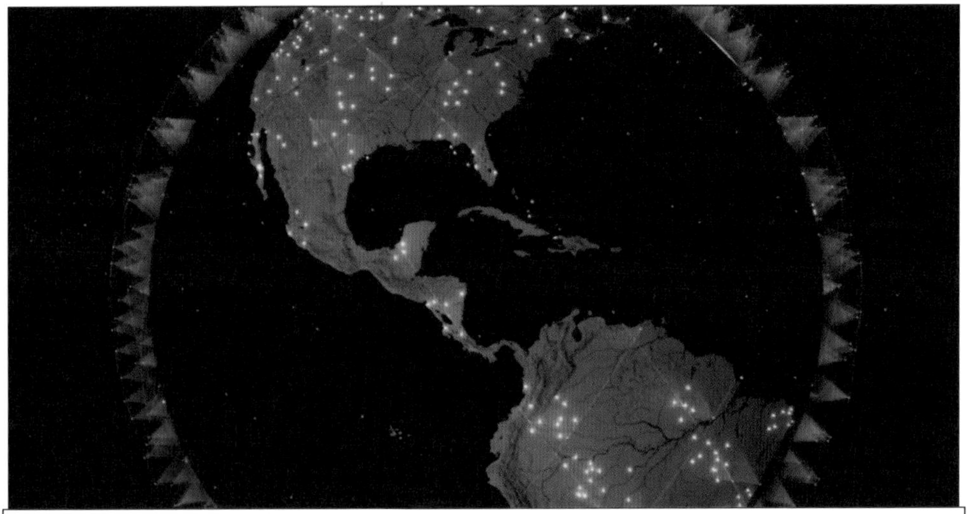

Starlink internet constellation, credit SpaceX

SpaceX plan to launch an initial constellation of at least 1,000 Ka/Ku band satellites into LEO (Low Earth Orbit) to provide affordable internet access around the world(26). This low orbit should help reduce latency (satellite delay) to less than 20 milliseconds for local transfers, effectively making response instantaneous from a human perspective. In May 2019 SpaceX successfully launched 60 test satellites to prove ground communications and largescale deployment. SpaceX plan to use flight proven Falcon 9s to launch up to 4,425 internet satellites by 2024, with another 7,518 to follow by 2027 in order to maximize constellation capacity(27).

Laser communication will also be used between satellites, which promises superior performance to conventional fiber optic networks for long distance data transfer. The eventual goal is to move the internet to space, allowing much faster, more reliable and secure communications around the world.

"In the long term it will be like rebuilding the internet in space. The goal will be to have a majority of long distance internet traffic going over this network – and about 10% of local and consumer and business traffic(28)." ~ *Elon Musk*

The idea of constructing an internet constellation in LEO dates back almost to the birth of the internet. Over the years many have pursued this vision but were unable to finance the staggering number of satellites and launch vehicles required to realize their dream. However, SpaceX's original approach provides some unique advantages to help overcome these problems, for example: -

60 Starlink satellites prepared for launch, credit SpaceX

- **High Density Deployment** – SpaceX intend to accelerate the rate of deployment by launching large numbers of Starlink satellites on each vehicle

Launch Vehicle	Satellite Capacity
Falcon 9 Block V	60
Starship	300+

- **Rocket Reusability** – SpaceX vehicles are designed for rapid reuse, so capable of sustaining the high launch cadence required to deploy and maintain such large constellations. In addition, these flights are relatively inexpensive, little more than the cost of a disposable second stage for Falcon 9 – or the cost of fuel for the fully reusable Starship

- **Vertical Integration** – all launch vehicles and satellites are built inhouse at SpaceX, significantly lowering final cost

- **Subsidized Launch** – Starlink satellites are comparatively light (~227kg) and surprisingly compact, hence other commercial satellites can be added to each flight, effectively reducing the cost to launch Starlink

- **Early Revenue** – only 1,000 satellites are required for the constellation to enter service, allowing further expansion to be self-funding

- **Google Support** – Google seem to appreciate the SpaceX approach and have already invested $900m in Starlink. If funding becomes an issue for completing this project, Google has some very deep pockets

"He [Elon Musk] wants to go to Mars. That's a worthy goal... I'd like for us to help out more than we are(29)." ~ *Larry Page, CEO of Alphabet Inc. (parent company for Google)*

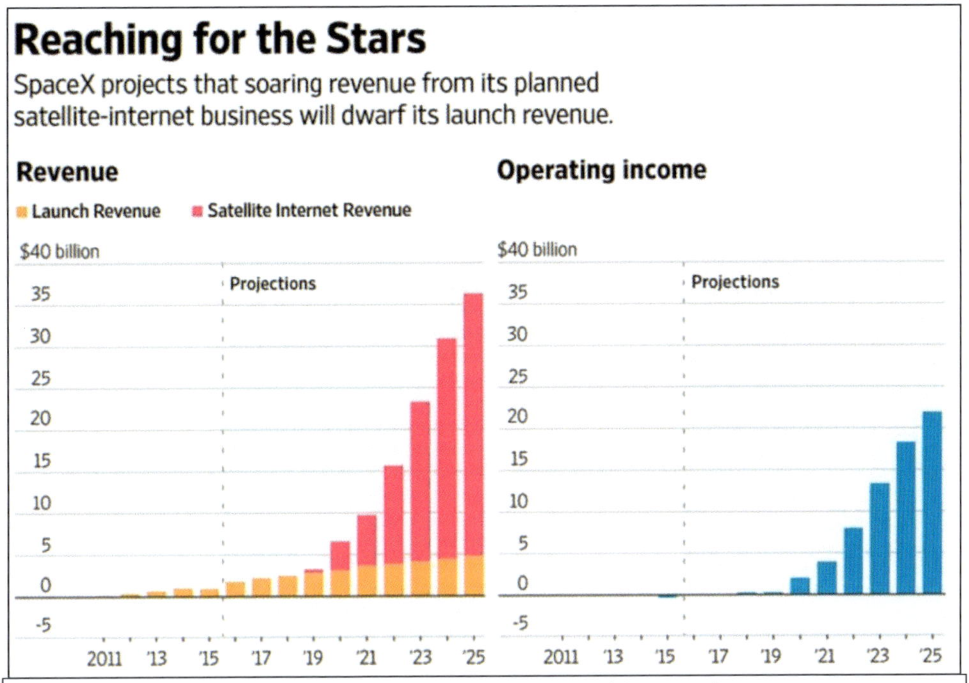

SpaceX projected revenue and return (circa 2015)

The potential revenue from supplying broadband to the world is staggering. It seems likely the construction of over 1,000 satellites and ground support stations could consume all of SpaceX's cash reserves and devour the lion's share of their launch revenue in the short term. However, in a few years' time, after system rollout, the return from internet satellites should swing hard in SpaceX's favor. In fact, they project overall revenue should exceed $35bn p.a. by 2025, the majority of which derives from their internet satellite business.

"We can do it, no question. We can fund both developments [Starlink and Starship], depending on the time frame you're talking about. But Elon is impatient to get to Mars, so we'll have to get a bit creative with the financing(30)." ~ *Gwynne Shotwell*

The final return from this internet constellation should depend on how the service is priced, whether there are multiple LEO constellations in close competition and how existing suppliers respond to new entrant(s). However, competition may become less of a problem long-term, as it has been revealed SpaceX intend to launch an additional 30,000 satellites (following the initial 12,000), to further improve bandwidth and global coverage(31). No doubt a return of hundreds of billions p.a. could eventually be realized, considering the potential market is every person, business and AI on the planet

MILITARY SUPPORT

US Congress want to form a military 'Space Corp'

In the near future, the US Military are looking to expand space activities, either through the formation of a "Space Corp" service within the Air Force or an entirely new branch of the military called "Space Force." No doubt the DoD miss certain capabilities provided by the Space Shuttle, which SpaceX's new Starship rocket should exceed in all categories (crew launch capability, satellite retrieval, in space servicing, reusability etc), making it a strong candidate for future DoD work.

"Let's say you have a satellite and you launch and something goes wrong... Starship has a capability to open its payload bay, either bring the satellite back in, close it, pressurize it, work on it and redeploy it. If you want to go see how your satellite is doing and if you're getting interference in the GEO belt, maybe you want to go up there and take a look at your neighbors, seeing if they're cheating or not, Starship will basically allow people to work and live in space and deploy technology that has not been able to be deployed(32)." ~ *Gwynne Shotwell*

Certainly the US military have shown increased interest in SpaceX's plans for Starship development, even going so far as to help fund a prototype Raptor engine(33). However, if the military wish to employ Starship in any capacity, they would need to offer reasonable remuneration in advance to secure these services. Depending on the complexity of the mission, they might easily pay hundreds of millions to adapt Starship to their mission requirements, with a comparable amount to cover the cost of each operational flight. Unfortunately, military procurement and Congress are relatively slow, so SpaceX might have to wait until at least 2020 for any additional funding, assuming the military view Starship as a suitable route to achieve space supremacy.

Starship prototype construction at Boca Chica Texas, credit SpaceX

So far, SpaceX have managed to finance Starship development without federal support. This allows them complete control of the design process, which means they can build Starship in the minimum time possible, fully optimized for their future requirements.

"You need to [try to not] get money from the government, otherwise the government will tell you what to build and how to build it... they will tell you how to build this and that's just not always – I mean for some things it's the best to do, but in others it's actually not(34)." ~ *Hans Koenigsmann, Vice President of Build and Flight Reliability at SpaceX*

While this refusal to receive government money might seem extreme, no doubt when SpaceX are happy with the Starship design, they will offer it to all suitable customers; military, civil and commercial. Likely it will become the workhorse spacecraft for cislunar space and help fund further development for Mars missions.

CONTINGENCY FUNDING

If SpaceX need to improve their finances they could tap venture capital funds at relatively short notice. In January 2015 Google and Fidelity were allowed to purchase a ten percent stake of SpaceX(35) in a private transaction. The company was valued at $10bn at the time, so effectively SpaceX netted $1bn in new finance. Steve Jurvetson, (one of SpaceX's prior investors) also wanted to improve his stake but his offer was politely declined. It seems $1bn was sufficient to cover the initial development cost of their internet satellites, which Google were keen to pursue.

It should be noted SpaceX were valued at $1.3bn in February 2012(36), meaning their value increased nearly eight-fold in less than three years. In May 2019 SpaceX were valued at $33.3bn(37) after they raised a further $1.02bn of private investment to support Starlink development. Their current value is difficult to appraise, however, given their increasing activity and dominance in the expanding space market they seem destined to become the most valuable private company in existence. Given their rapidly escalating value and constant demand from investors, venture capital could provide SpaceX with considerably more funds to fulfill their long-term goals if needed.

"There is an unlimited amount of funding that the company could probably access globally in private markets... Everywhere I travel around the world, investors of all types – individuals, family offices, hedge funds, sovereign wealth funds or private equity – want to get into SpaceX(38)"
~ Robert Hilmer, head of business development at private market analysis group Equidate.

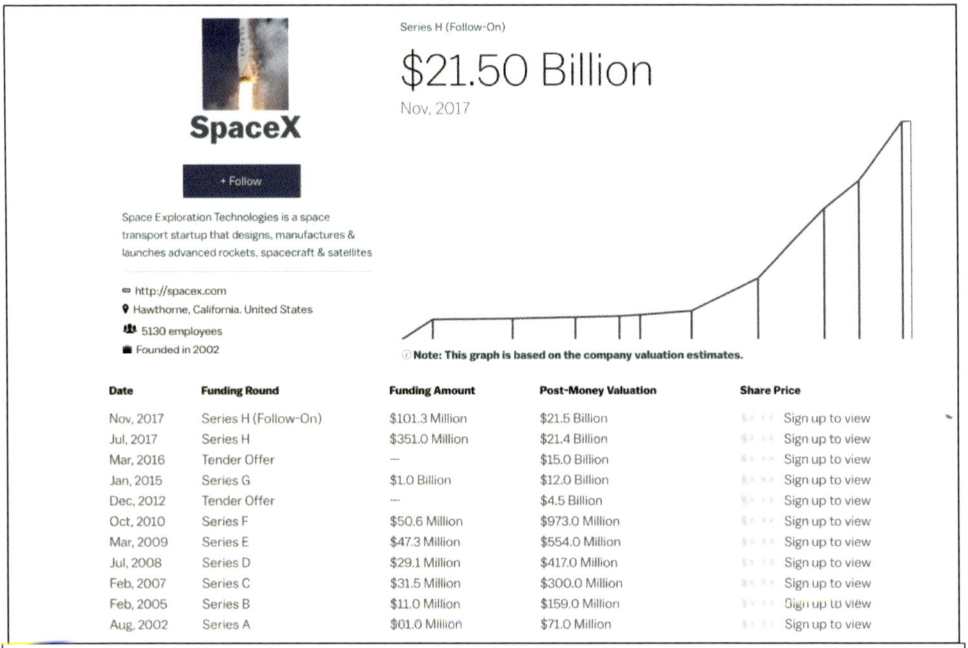

Date	Funding Round	Funding Amount	Post-Money Valuation	Share Price
Nov, 2017	Series H (Follow-On)	$101.3 Million	$21.5 Billion	Sign up to view
Jul, 2017	Series H	$351.0 Million	$21.4 Billion	Sign up to view
Mar, 2016	Tender Offer	—	$15.0 Billion	Sign up to view
Jan, 2015	Series G	$1.0 Billion	$12.0 Billion	Sign up to view
Dec, 2012	Tender Offer	—	$4.5 Billion	Sign up to view
Oct, 2010	Series F	$50.6 Million	$973.0 Million	Sign up to view
Mar, 2009	Series E	$47.3 Million	$554.0 Million	Sign up to view
Jul, 2008	Series D	$29.1 Million	$417.0 Million	Sign up to view
Feb, 2007	Series C	$31.5 Million	$300.0 Million	Sign up to view
Feb, 2005	Series B	$11.0 Million	$159.0 Million	Sign up to view
Aug, 2002	Series A	$01.0 Million	$71.0 Million	Sign up to view

SpaceX valuation chart and funding data, circa 2017

PUBLIC/PRIVATE FUNDING

"Ultimately this is going to be a huge public-private partnership. That's how the United States was established, and many other countries around the world. I think that's how it will happen. Right now we're trying to make as much progress as we can with the resources that we have available and just sort of keep moving both [commercial operations and Mars colonization] forward and hopefully I think as we show that this is possible, that this dream is real, not just a dream but that it is something that is being made real, I think the support will snowball over time(39)."
~ Elon Musk/IAC 2016

If SpaceX provide mass transport to Mars, passenger berths could be sold to space agencies at premium rates, potentially adding another substantial revenue stream to their portfolio. To illustrate, NASA's Curiosity Rover cost roughly $2.5

billion(40), hence SpaceX could realistically charge NASA a comparable amount to transport scientists to Mars for a two year sojourn.

It is customary to pay a substantial deposit to reserve commercial launch services hence SpaceX could reasonably ask NASA for $1bn per passenger, many years in advance of the first Mars expedition. It's possible NASA might decide not to purchase these passenger berths because they judge they are too far along with their own Mars plans. However, that should allow SpaceX to offer these places to other space agencies, who would likely jump at the chance to explore Mars.

Mars Curiosity Rover, credit NASA/JPL

Apparently Starship does figure in NASA's plans for the future, although details are scant. No doubt more will be revealed as Starship flights unfold.

"Starship is a really big vehicle. Being able to refuel it will be necessary to become a vehicle that can get to the moon. SpaceX can use it for their reasons and we can use it for our reasons(41)."
~ *Jim Bridenstine/NASA Administrator*

Private citizens interested in going to Mars can also contribute to the funding effort. SpaceX intend to allow commercial passengers to reserve berths on Starship after they prove Earth reentry(42). Note these reservations will likely be placed many years in advance of when Mars flights are possible, allowing SpaceX to accumulate considerable funds to assist with development (similar to cislunar tourism flights).

Generally, as the amount of public and private money invested in Mars increases this should attract more private capital. For example, private investors could supply Bigelow habitats, Mars internet infrastructure or independent ISRU ventures. SpaceX need only supply the transport to start the ball rolling; as the say: "build it and they will come."

CONCLUSIONS

1. SpaceX could finance development of a Mars transport vehicle solely from launch revenue, even if that revenue proves significantly less than predicted.

2. SpaceX finances will be tight in the next five years if they pursue their Mars colonization and internet satellite projects in parallel. If flight rates are reduced, some additional funding (e.g. venture capital etc) might be required.

3. If SpaceX manage to build their LEO internet constellation, it's possible they could independently finance the construction of a city on Mars (using majority in situ resources). Elon Musk stated at the opening of the SpaceX Seattle Office: -

"This [LEO internet constellation] is intended to generate a significant amount of revenue and help fund a city on Mars(28)."

4. Any long stay mission on Mars will likely attract both private and public funding. Some form of public-private partnership should begin after SpaceX demonstrate their Mars rocket architecture, possibly sooner.

Alcubierre Interstellar Spacecraft

5. Currently SpaceX's value increases by around $1bn per month. In the early 2020's, after the first Starship launch, their value should increase by $1bn per week. If SpaceX reach Mars, their value might increase $1bn per day. Should they subsequently decide to go public, the sale of these shares could provide astronomic capital (possibly making them the first multi-trillion dollar company), allowing SpaceX to finance some even more ambitious project(s).

"So as soon as we've got a base on Mars, we can see a base on the moon but certainly one on Mars, which creates a very powerful forcing function for making space technology better every year and that is what will lead us to interstellar travel(39)." ~ *Elon Musk*

6. SpaceX are going places, both lit. and fig.

[1]https://twitter.com/elonmusk/status/995462943079723008
[2]https://gist.github.com/theinternetftw/5ba82bd5f4099934fa0556b9d09c123e
[3]http://spacenews.com/spacexs-low-cost-won-gps-3-launch-air-force-says/
[4]https://www.spacex.com/about/capabilities
[5]https://spaceflightnow.com/2018/03/01/rideshare-mission-for-u-s-military-confirmed-as-second-falcon-heavy-launch/
[6]http://spacenews.com/spacex-aims-to-debut-new-version-of-falcon-9-this-summer/
[7]http://spacenews.com/spacex-wins-5-new-space-station-cargo-missions-in-nasa-contract-estimated-at-700-million/
[8]https://oig.nasa.gov/docs/IG-18-016.pdf
[9]https://spacenews.com/spacex-president-gwynne-shotwell-we-would-launch-a-weapon-to-defend-the-u-s/
[10]https://blogs.nasa.gov/bolden/2014/09/16/american-companies-selected-to-return-astronaut-launches-to-american-soil/
[11]https://youtu.be/DaJ0n0j-UB8?t=3107
[12]http://bigelowaerospace.com/b330/
[13]https://youtu.be/uBaLYDbk4fY?t=171
[14]http://www.thespacereview.com/article/1719/1
[15]https://www.nasa.gov/nextstep/freeflyer
[16]https://www.reddit.com/r/spacex/comments/9ggtma/Starship_manned_moon_mission_thread_livestream_at/e66pddw/
[17]https://arstechnica.com/science/2017/03/for-the-first-time-nasa-has-begun-detailing-its-deep-space-exploration-plans/
[18]https://twitter.com/jeff_foust/status/847516225081159681
[19]http://spacenews.com/the-big-changes-that-may-not-be-coming-to-nasa/
[20]https://arstechnica.com/science/2016/08/how-much-will-sls-and-orion-cost-to-fly-finally-some-answers/
[21]https://spaceflightnow.com/2017/03/31/spacex-flies-rocket-for-second-time-in-historic-test-of-cost-cutting-technology/
[22]https://youtu.be/2mCGbguCw2U?t=971
[23]http://www.spacex.com/press/2012/12/19/spacex-falcon-9-upper-stage-engine-successfully-completes-full-mission-duration
[24]https://youtu.be/uBaLYDbk4fY?t=1254
[25]https://futurism.com/elon-musks-spacex-just-announced-hundreds-of-open-positions/
[26]https://spacenews.com/musk-says-starlink-economically-viable-with-around-1000-satellites/
[27]https://spacenews.com/fcc-oks-lower-orbit-for-some-starlink-satellites/
[28]https://youtu.be/AHeZHyOnsm4?t=794
[29]http://www.businessinsider.com/larry-page-elon-musk-2014-3?IR=T
[30]https://www.reddit.com/r/spacex/comments/75ufq9/interesting_items_from_gwynne_shotwells_talk_at/do94zn3/
[31] https://twitter.com/CHenry_SN/status/1184139048111226886
[32]https://youtu.be/qWPaopcU_hE?t=431
[33]https://spacenews.com/air-force-adds-more-than-40-million-to-spacex-engine-contract/
[34]https://www.teslarati.com/spacex-executive-nobody-paid-us-to-make-falcon-heavy/

[35]http://www.space.com/28316-spacex-elon-musk-google-fidelity-investment.html
[36]http://www.cnbc.com/id/47207833
[37]https://www.cnbc.com/2019/05/31/spacex-valuation-33point3-billion-after-starlink-satellites-fundraising.html
[38]https://www.cnbc.com/2018/04/13/equidate-spacex-27-billion-valuation-shows-unlimited-private-funding-available.html
[39]http://diyhpl.us/wiki/transcripts/spacex/elon-musk-making-humans-a-multiplanetary-species/
[40]http://www.space.com/10762-nasa-mars-rover-overbudget.html
[41]https://eu.floridatoday.com/story/tech/science/space/2019/10/12/nasa-shows-interest-spacexs-starship-orbital-refueling-ambitions/3957775002/
[42]https://twitter.com/elonmusk/status/1142902063203942400

Chapter 6: How SpaceX Can Travel to Mars

Starship Launch System, credit SpaceX

It is a very long way to Mars, on average its 225 million kilometers(1) from Earth. However, every two years the two planets come closer, due to their differing orbital speeds around the sun. Yet even when Earth and Mars are in conjunction (i.e. adjacent), the gulf between them is still vast, varying from 50 to over a 100 million km. The Apollo astronauts, by contrast, travelled 'only' a third of a million kilometers to reach the moon. Hence to reach Mars in any reasonable amount of time will require a monstrously large rocket, more powerful even than the mighty Saturn V moon rocket.

SpaceX terminology: The Starship Launch System has gone through many design iterations, with each iteration bearing a different name. This Mars capable rocket has previously been called: "BFR" (Big 'Falcon' Rocket), "ITS" (Interplanetary Transport System) and "MCT" (Mars Colonial Transporter). SpaceX currently use the term "Starship" for both the upper stage and full stack rocket – and recently "Starship Launch System" to describe the entire vehicle. For clarity, I will use Starship to refer to the upper stage and Starship Launch System (or Starship LS) for the full stack rocket, as we continue to explore this extraordinary vehicle.

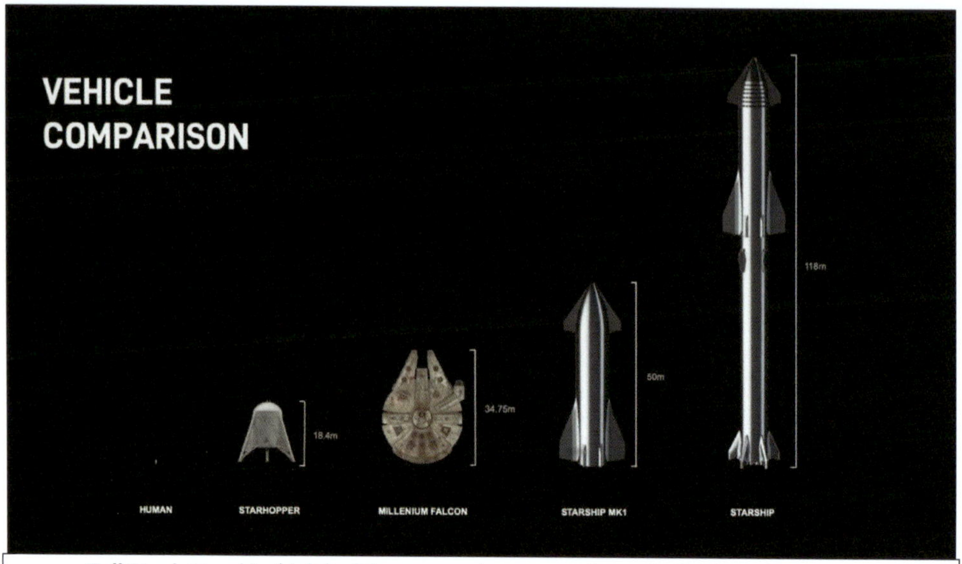

Full Stack Starship (right) with average size person for scale (left), credit SpaceX

Yes, SpaceX are building an enormously powerful launch vehicle which they affectionately call the "Starship Launch System." In fact, this launch vehicle is so capable it could be used to explore and colonize the entire solar system. Details of this extraordinary SHL (Super Heavy Lift) rocket were announced by Elon Musk at IAC 2017(2), with some additional design details provided by Elon during the Lunar Mission presentation in 2018(3) and Boca Chica update, in 2019(4). SpaceX predict Starship LS should be capable of transporting over 100 mt (metric tons) of cargo or 100 people to the surface of Mars. This figure seems high, however, they intend to establish a colony on Mars which means they need to take everything to survive until the next transports arrive (ideally in two years). Starship LS and Saturn V are both designed to transport people beyond Earth orbit, so let's take a look under the hood to see how far rocket science has progressed in the intervening period.

Launch Vehicle Comparison Table: Starship LS vs Saturn V

	Starship LS	Saturn V	Ratio
Gross lift-off mass (mt)	5,000	3,039	1.6
Lift-off thrust (MN)	72	35	2.06
Lift-off thrust (mt)	7,400	3,579	2.06
Vehicle height (m)	118	111	1.06
Tank Diameter (m)	9	10	0.9
Expendable LEO Payload (mt)	300(5)	135	2.22
Reusable LEO Payload (mt)	150	0	∞

While Starship LS and Saturn V are comparable in size, Starship LS performance far exceeds the mighty Saturn V, both in thrust and payload to LEO. More significantly Starship LS is reusable (with all the cost savings and capabilities that implies), a huge advance over the expendable Saturn V design, which couldn't be described as a sustainable mode of transport. It should be noted of course that orbital rocketry was barely out of its infancy when Saturn V was designed. Starship LS benefits enormously from new materials like cryogenically formed stainless steel (capable of withstanding deep cryo and high reentry temperatures) and Raptor full flow staged combustion engines (which utilize a broadly superior methalox system). The Saturn V F-1 engines were brutally powerful but the 37 Raptors, used to launch Starship LS, promise an altogether smoother ride.

In the final analysis Saturn V was a Moon mission rocket, whereas Starship LS is designed to colonize Mars (which can then be used as a staging post to colonize the greater solar system) – a much more challenging proposition.

ITS Spacecraft transits Jupiter, credit SpaceX

Both rockets rely on chemical propulsion, yet the way they use it is profoundly different, so let's break down Starship LS further to see where all this world spanning power is coming from.

Super Heavy (Booster)

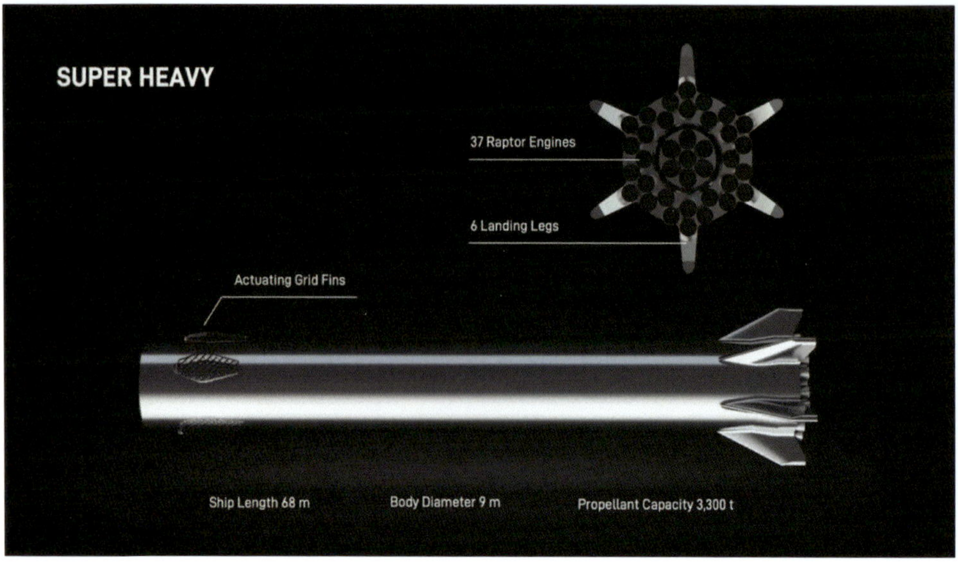

Length	68 m
Diameter	9 m
Dry Mass	380 mt
Propellant Mass	3,300 mt
Raptor Engines	37
Sea Level Thrust	72 MN
Fabrication Cost	~$230m*
Total Launches	~1000
Launches Required	5/Mars trip
Maintenance Cost	~$0.2m av./flight*
Launch site Costs	~$0.2m/flight*
Total Cost	~$11m/Mars trip*

The Super Heavy booster will accelerate the Starship second stage to ~8,000 km/hour before separating. Super Heavy then reverses direction and boosts back toward the launch site. Four diamond shaped grid fins are deployed to help guide the booster through the atmosphere and back to the launch center. Immediately before touchdown six landing legs extend from the base of the booster, allowing it to land with precision next to the launch pad. Only 7% of the initial propellant load is required to perform the boost-back and landing maneuvers. Super Heavy is designed to be reused 1000 times and can withstand 20 g's of acceleration - possibly 30 or 40 g's without breaking up(6).

*These are fairly optimistic cost estimates for the precursor ITS design which can be used as guidelines for the slightly smaller and less expensive Starship LS.

Starship (Upper Stage)

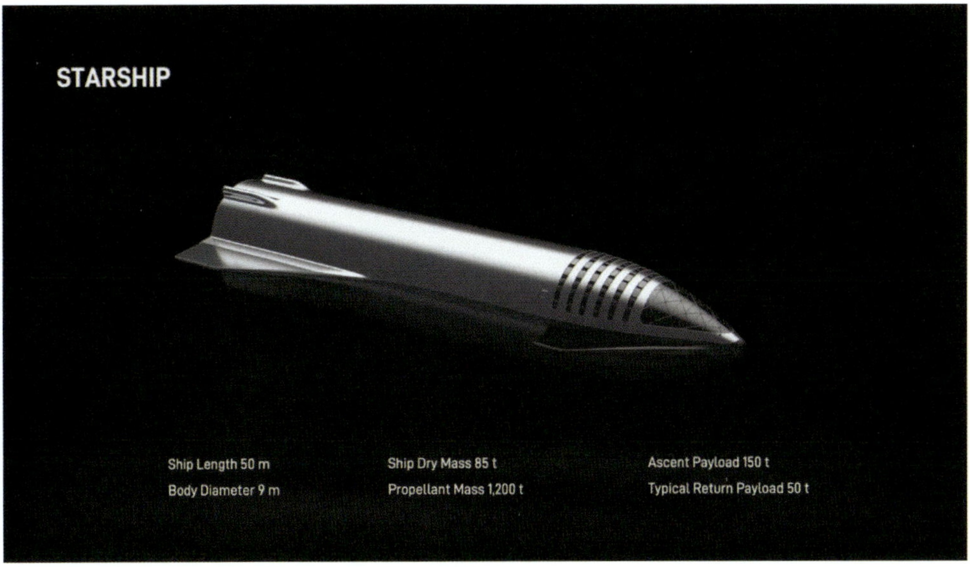

Length	50 m
Max Diameter	9 m
Dry Mass	105(7) mt
Propellant Mass	1,200 mt
Ascent Payload to LEO	150 mt
Max Payload (Disposable)	300 mt
Return Payload (from Mars)	50 mt
Passengers	100
Raptor Engines	3 Sea Level 3 Vacuum
Sea Level Thrust	5.58 MN
Vacuum Thrust	6 MN
Fabrication Cost	~$200m*
Total Launches (to Mars)	12
Launches Required	1
Maintenance Cost	~$10m*
Cost/Tons to Mars	~$140,000*
Total Cost	~43m/Mars mission*

An advanced heat shield is fitted across Starship's lower face to facilitate aerocapture (decelerating to orbital speed) and EDL (Entry, Descent and Landing). Two pairs of actuated fins, fitted forward and aft, are used for atmospheric flight control. As the vehicle prepares for vertical propulsive landing, these fins are folded into the body to minimize any aerodynamic interference. Then six landing legs extend from the base of the vehicle to facilitate smooth landing on unprepared surfaces. Starship is designed to transport up to 100 passengers, with on board amenities including personal cabins, a large communal area and galley.

*These are fairly optimistic cost estimates for the precursor ITS design which can be used as guidelines for the slightly smaller and less expensive Starship LS.

Starship Tanker

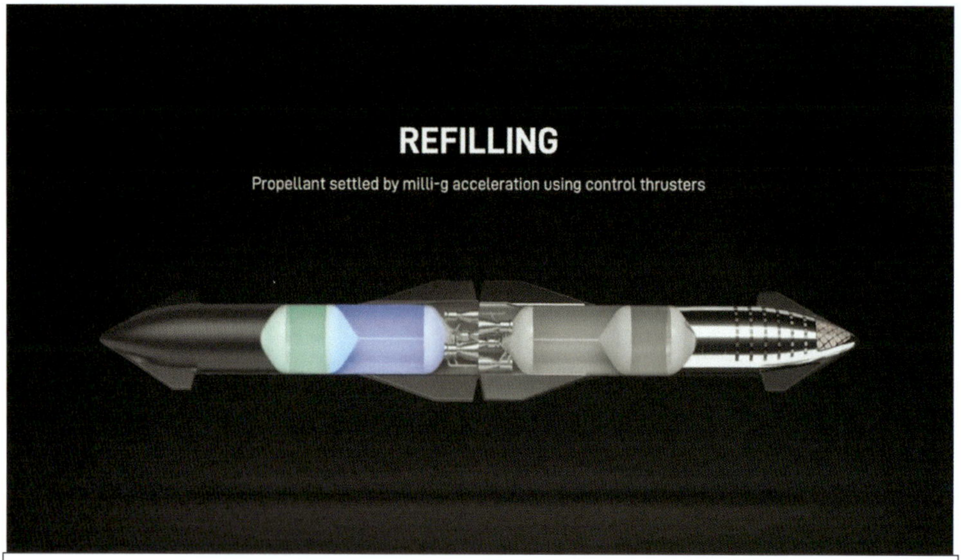

Tanker (left), refueling Starship (right), credit SpaceX

SpaceX also intend to launch a Starship Tanker with the Super Heavy booster. This Tanker will be used to transport propellant to LEO, in order to refuel a colony Starship in orbit. Specifications for the Tanker Stage should be similar to Starship, except as follows: -

Dry Mass	~50 mt
Propellant Mass	1,100 mt
Propellant to LEO	~220 mt
Fabrication Cost	~$130m*
Total Launches	100
Launches Required	4/Mars trip
Maintenance Cost	~$0.5m*
Total Cost	~$8m/Mars trip*

It should be noted these are the figures for version 1.0 Starship. Elon Musk has suggested version 2.0 would have an 18m diameter(8) which implies 4 times greater lifting capacity. To illustrate this capability: Tanker v2.0 could fully refuel a v1.0 Starship in LEO after a single launch.

*These are fairly optimistic cost estimates for the precursor ITS design which can be used as guidelines for the slightly smaller and less expensive Starship LS.

Raptor Engine

Production Raptor on test stand at McGregor Texas, credit SpaceX

"FFSC [Full Flow Stage Combustion] is the ultimate architecture for converting propellant into rocket velocity(9)." ~ *Elon Musk*

Engine Cycle	Full-Flow Staged Combustion	
Oxidizer	Subcooled Liquid Oxygen	
Fuel	Subcooled Liquid Methane	
Chamber Pressure	300 Bar(10)	
Throttle	20-100% Thrust	
Raptor Variant	Sea Level	Vacuum
Exit Diameter	1.3 m	~2.8 m(11)
Specific Impulse (Isp)	330 s	380 s
Thrust	1.96 MN	2.16 MN

Raptor should produce over twice the thrust of the existing Merlin 1D engine, despite having a comparable physical size. This similarity of size should allow the SpaceX production line to smoothly transition from Merlin to Raptor engines. They intend to produce 2 Raptors per day(12) at a relatively low cost of $200,000 per engine(13).

However, SpaceX always have an eye for the next iteration, as Elon revealed: "Planning on a simplifying mod to Raptor for max thrust, but no throttling, to get to 250 mt [~2.45 MN] level(14)." Likely this Raptor-Max would be used initially on the Super Heavy booster, where raw power is needed most, alongside the seven throttleable Raptors required for landing.

Engine Design

Production Raptor engine test, credit SpaceX

Starship LS will use Full Flow Staged Combustion 'Raptor' engines on both rocket stages. This type of engine maximizes the efficiency of the combustion process, ensuring most of the energy released is converted into usable thrust. Overall Raptor promises to be many times more durable than conventional engines, increasing the number of times it can be fired and burn durations. Perhaps the most highly stressed component of the engine is the turbopump, so many of the improvements are focused on this key component. Twin turbopumps are used to pump propellant at prodigious rates into the combustion chamber, so they could be thought of as the beating heart of the engine. Essentially anything which reduces stress inside these turbopumps should improve engine longevity. Here's a list of features used in the engine's design to improve engine durability: -

"Raptor is designed for ~1000 flights with negligible maintenance." ~ *Elon Musk*

1. Fuel and oxygen propellants have independent turbopumps, hence eliminating the fuel-oxidizer turbine interseal(15), which is a known point of failure in traditional turbopump designs.

2. Preburners are integral to the turbopump, eliminating any need for external power transmission components.

3. Fuel and oxidizer are designed to be mixed in the propellant flow, eliminating the need for elaborate seals between the pump and preburner (used on prior engines).

4. Flow rate through FFSC turbopumps is a lot higher, allowing them to run cooler and at lower pressure than conventional turbopumps.

5. Hydrostatic bearings (which support moving parts on a thin film of pressurized fluid) effectively eliminate wear and tear on load bearing components.

6. Deep cryo propellant produces less cavitation and improves turbopump cooling. This should reduce turbine stress and result in more reliable operation.

7. High temperature superalloys have been developed at SpaceX's internal foundry.

"SpaceX metallurgy team developed SX500 superalloy for 12000 psi, hot oxygen-rich gas. It was hard. Almost any metal turns into a flare in those conditions." ~ *Elon Musk*

8. Spark ignited dual redundant torches (using gaseous propellant) provide smoother and more reliable engine startup.

9. Early engines were driven to their limit on the test stand to discover where they would fail. Then the following engine was improved based on lessons learnt – and the process repeated.

"The max chamber pressure run damaged Raptor SN 1 (as expected). A lot of the parts are fine for reuse, but next tests will be with SN 2, which is almost done...SN 2 has changes that should help" ~ *Elon Musk*

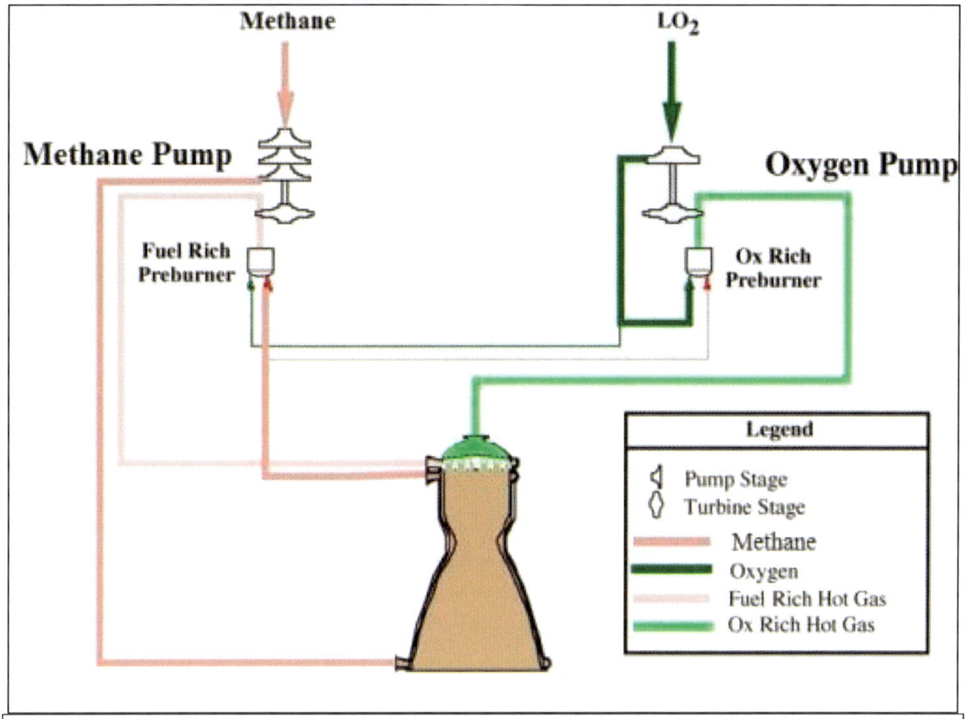

Methalox FFSC Engine Schematic

However, SpaceX didn't stop there, in fact they have taken a holistic approach to designing the ideal engine for Mars missions, including the use of fuel. Raptor uses an unusual combination of propellants: liquid methane and oxygen (LOX), in what is called a methalox system. The main advantages of using methalox are: -

1. Methane (when derived from LNG/Liquid Natural Gas) is relatively inexpensive compared to more conventional kerosene fuel. In addition methalox engines require a higher proportion of oxygen to methane than a comparable kerolox system (methalox 3.5 : 1 vs kerolox 2.5 : 1). This further reduces the propellant cost because oxygen is very inexpensive, even cheaper than methane.

2. Methane causes less coking (accumulation of carbon) in the engine compared to the more conventional kerosene, which makes for a more reliable performance over multiple launches.

3. Methane and oxygen can be autogenous, in other words they don't require helium tank pressurization equipment (which reduces weight, eliminates points of failure and increases the volume available for propellant).

4. Both methane and oxygen can be manufactured in situ on Mars, hence removing the need to carry extra propellant for the return journey, which significantly improves the payload delivered to Mars.

5. Propellants will be cooled to deep cryo temperatures to make them denser, which allows more propellant to be stored in the same volume tanks.

6. Cryogenic methane and LOX have similar temperatures, hence reducing thermal insulation requirements between the fuel and oxidizer tanks.

7. Methalox is a relatively fast combustion process, hence can provide higher chamber temperature and pressure, which improves overall engine efficiency (Isp). Fast combustion also allows the combustion chamber to be smaller, which should reduce the overall engine weight.

STARSHIP LS CONSTRUCTION

Starship assembly at Boca Chica Texas, credit SpaceX

The Starship LS external skin largely supports the structure of the vehicle in what's called a monocoque design. The external skin also doubles as the wall for the propellant tanks to save weight and cost. It has been reported they intend to construct the entire airframe from stainless steel, for both the Super Heavy booster and Starship upper stage. The skin of the vehicle should be highly polished for aesthetic reasons and to reduce heat absorption during fuel loading – allowing them to dispense with thermal insulation.

Material Change

Originally SpaceX intended to build the BFR airframe from advanced carbon fiber composites, to improve structural strength and reduce overall weight. However, in practice carbon fiber proved expensive and difficult to work, with some inherent quality issues. So SpaceX switched to a more conventional material, stainless steel (reportedly a special variant of 301 stainless is used for both the Super Heavy booster and Starship upper stage). Overall stainless steel is far cheaper to manufacture and easier to maintain – an important consideration for Moon or Mars based operations.

"Yes. Very easy to work with steel. Oh, and I forgot to mention: The carbon fiber is $135 a kilogram, 35 percent scrap, so you're starting to approach almost $200 a kilogram. The steel is $3 a kilogram(16)." ~ *Elon Musk/Popular Mechanics*

This decision appears counter-intuitive because stainless steel is normally much heavier than carbon fiber and might add extra weight to the vehicle. However, stainless has better strength-to-weight performance at cryogenic temperature and is vastly better at high temperature. Hence while carbon fiber is stronger at room temperature, stainless proved superior for the extreme range of temperatures experienced by the vehicle during operation e.g. extreme cold when loaded with cryogenic fuel and extreme heat during atmospheric braking. Space too is a rugged environment, it is either extremely hot (when exposed to direct sunlight) or extremely cold (if in shadow) so from the engineer's point of view, stainless was an obvious selection – from a space perspective.

It's worth noting, a previous advanced launch vehicle, Lockheed Martin's X-33 VentureStar, encountered similar problems using carbon fiber but was unable to transition to metal, which eventually caused the project to founder. Hence this material change can be seen as an important advance, which allowed SpaceX to quickly proceed to the next stage of development – flight tests of a Starship prototype (called Starhopper) in early 2019.

Advanced Heat Shield

"Thin [heat resistant] tiles on windward side of ship & nothing on leeward or anywhere on booster looks like lightest option(17)." ~ *Elon Musk*

Conventional heat shield materials are designed to ablate, i.e. slowly evaporate, under fierce reentry temperatures. This produces a layer of vapor which helps insulate the vehicle from the extreme heat, at the cost of some heat shield thickness and degradation in structure. However, Starship will use ceramic (non-ablative) heat tiles across 40% of the hull (on the windward side) to counter heating effects during atmospheric entry. Ideally the remaining stainless steel hull will be left untreated, except for a little polish. This dual approach to heat protection promises many advantages over an all-enveloping heat shield design: -

1. Stainless steel is much lower cost, easier to manufacture/maintain and results in a lighter airframe than one covered with inherently bulky ablative materials.

2. Ceramic heat tiles can be relatively thin because the underlying stainless steel is highly durable.

3. The hexagon shaped heat tiles (which primarily consist of strands of glass) should give the vehicle greater rigidity, essentially acting as a spine. This should

allow thinner section stainless steel to be used across the rest of the craft, reducing overall vehicle mass.

4. The heat protection system shouldn't degrade, allowing numerous atmospheric entries at interplanetary velocity.

5. Any excess heat on the windward side can be conducted to the remaining 60% of the hull, allowing it to be used as a heat sink/radiator.

STARSHIP LS DESIGN CONCLUSIONS

1. Methane propellant has been largely overlooked because it has lower density than more conventional kerosene hence requires larger volume tanks to contain the same mass of propellant. Methalox engines have a higher Isp (i.e. performance) compared to kerolox but that benefit is offset by the larger tank mass required, so kerosene is normally used because it is easier to handle. However, deep cryo methalox promises to deliver the benefits of higher Isp with minimal increase in tank volume.

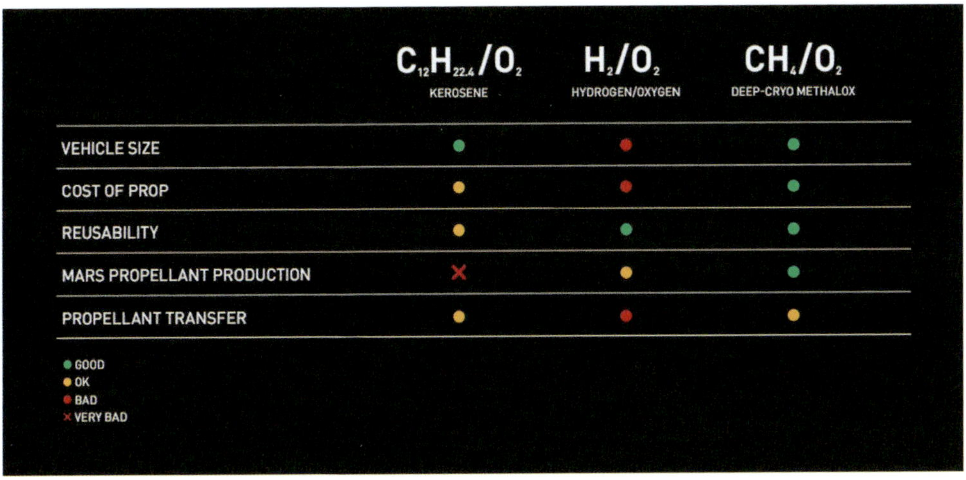
$C_{12}H_{22.4}/O_2$
KEROSENE
H_2/O_2
HYDROGEN/OXYGEN
CH_4/O_2
DEEP-CRYO METHALOX
VEHICLE SIZE
COST OF PROP
REUSABILITY
MARS PROPELLANT PRODUCTION
PROPELLANT TRANSFER
GOOD
OK
BAD
VERY BAD

2. Raptor lowers weight, has superior performance, is less expensive to run and potentially more reliable than conventional rocket engines, making it ideal for rapid stage reuse and deep space missions.

"Other rocket engines were designed for no (or almost no) reuse. Raptor is designed for heavy & immediate reuse, like an aircraft jet engine, with inspections required only after many flights, assuming instrumentation shows it good. Using hydrostatic bearings certainly helps(18)."
~ *Elon Musk*

3. The combination of a light airframe and powerful engines allows both stages to be reusable, i.e. each stage can be launched and landed many times. In addition, Starship can be launched from Mars to anywhere in the solar system and should only require refueling after landing at each new destination. Thus Super Heavy booster should only be required to launch from high gravity worlds, like Earth.

4. Judging by SpaceX's past performance, Starship LS architecture should rapidly evolve throughout the development process. Initial development could proceed relatively quickly because both stages are designed to be reusable. Each stage could be launched and landed separately, allowing the effect from each operation to be precisely gauged through direct inspection of the returned stages. Barring accidents, reusability should remove the need to manufacture numerous test vehicles, greatly accelerating the test program. In addition, Starship can be flown as a single stage, allowing it to be independently tested through a large regime of flight conditions.

STARSHIP LS OPERATION

BFR launch, credit SpaceX

Starship LS promises to become the most powerful rocket ever constructed, bar none. It far exceeds the Saturn V or N1 Moon rockets and should even surpass the stupendous Nova rocket – NASA's Mars rocket design from the 1960's.

Super Heavy Booster Operation

Super Heavy booster performs boost-back maneuver, credit SpaceX

To place 100 metric tons of useful payload on Mars will require something plucked from the pages of space opera. The bulk of this rocket will reside in its first stage, i.e. Super Heavy booster (3,680 mt, 68 m overall length). This booster will use 93% of its propellant to hurl the Starship second stage into space prior to separation. But unlike the disposable boosters of its predecessors, Super Heavy will then use the remaining propellant to reverse course and deploy four grid fins to guide its entry into the atmosphere and final approach, allowing it to land safely at its original launch site. This procedure is designed to maximize the use from each booster by minimizing the booster turnaround time.

Once the Super Heavy booster returns to the launch site, it is immediately inspected, refurbished and refueled, ready for the next flight. Hopefully within days the booster will be ready to launch a new second stage (the Starship tanker) into orbit before landing safely again at the same launch site. In other words, they intend to use the booster like a cannon to shoot multiple payloads into space, simply reload and fire. Starship LS is weather agnostic, which means it has been designed to launch in almost any weather conditions. The Mars window is relatively narrow, during which time SpaceX will need to launch as frequently as possible, so Starship LS's ability to virtually ignore any vagaries of climate is a great advantage.

Tanker is vertically integrated onto Super Heavy booster on launch pad, credit SpaceX

Ideally SpaceX want to operate two or more Starship LS in parallel from separate launch sites (Cape Canaveral Florida and Boca Chica Texas), which should allow one to be used for launch operations while the other is being inspected, refurbished and reintegrated with its new second stage payload. You could say they'll have one to wash and one to wear - except on a mega-scale. Operating two Super Heavy boosters would also provide redundancy, should a problem arise during launch, recovery or refurbishment.

Starship Spacecraft Operation

The entire upper stage of Starship LS is a combined space habitat and lander, called the Starship spacecraft. This spacecraft is designed to carry 100 metric tons of cargo (or a hundred people) to the surface of Mars. After separating from the Super Heavy booster, most of the upper stage propellant will be used to reach Low Earth Orbit (LEO). Then a Tanker will rendezvous with Starship, allowing it to be refueled on orbit. It will likely require three to five tanker flights to replenish the spacecraft's propellant tanks. This procedure significantly increases Starship's mass (to ~1,300 mt, most of which is propellant) which should greatly reduce the transit time to Mars compared to using an expendable SHL rocket.

Then when the planets align, Starship will use a series of burns to increase velocity until it achieves a high energy Mars transfer orbit. Interestingly they intend to use all Raptor engines, both vacuum and sea level, to perform these Mars transfer burns. The fixed vacuum engines will fire at maximum power, to supply the majority of thrust needed, while the Sea Level engines (which can be tilted in a process called gimballing) are used to control the direction of travel, by altering the angle of thrust.

"...it's actually good to fix high efficiency vacuum engines with giant nozzles in place & only thrust vector engines with smaller nozzles. Don't need a lot of room & moment of inertia is much lower... Yeah, gimbal SL engines at min throttle for control, so most of impulse goes through vac engines(19)." ~ *Elon Musk*

ITS Spacecraft enters Mars atmosphere, credit SpaceX

Then three to six months later the spacecraft enters the upper atmosphere of Mars and performs a high velocity (up to 7 km/s) aerobraking maneuver. Typically the vehicle will enter at 60 degree angle, which should provide sufficient lift to maintain altitude while progressively removing most of the horizontal velocity. When sufficiently slowed the spacecraft reverses its facing to perform a supersonic retro-propulsion burn, followed by propulsive landing. Entry deceleration will likely be high (peak 5 g), made possible by the high strength monocoque design.

This combination of aerobraking and supersonic retro-propulsion could provide many advantages over more conventional entry techniques. Aerobraking generates terrific heat as the spacecraft rams through the atmosphere, even when it's relatively diffuse, like for Mars. However, aerobraking is a tried technique and should provide an extremely efficient means to remove a significant proportion (over 99%) of the high entry velocity. As Starship enters the thicker lower atmosphere, the Supersonic Retro-Propulsive burn should expand the bow shock wave away from the engines, creating a virtual heat shield in front of the vertically descending spacecraft(20). In addition, by carefully throttling the engines this virtual heat shield can be shaped and angled, making it perform like a virtual wing, to better control the craft's descent trajectory. This should allow it to maintain a sufficiently high altitude to avoid terrain and safely bleed off any remaining entry velocity before landing precisely at the designated area.

After landing, the Starship spacecraft will be used as a habitat, until the cargo has been safely offloaded. Likely some form of pressurized structures will be erected to serve as surface habitats and/or greenhouses. Mars has all the resources necessary to make rocket propellant, hence methane and oxygen can be produced in situ using electrolysis and the Sabatier process. This In Situ Resource Utilization (ISRU) should allow the Starship spacecraft to be refueled and return to Earth without the need for staging or orbital refueling, thanks to Mars's low gravity, which is roughly one third of Earth's. For these reasons, ISRU will be vital to the Mars mission hence SpaceX are designing and developing the ISRU plant themselves. According to a recent report the ISRU design work is proceeding well(20), which probably means they began prototyping in 2018.

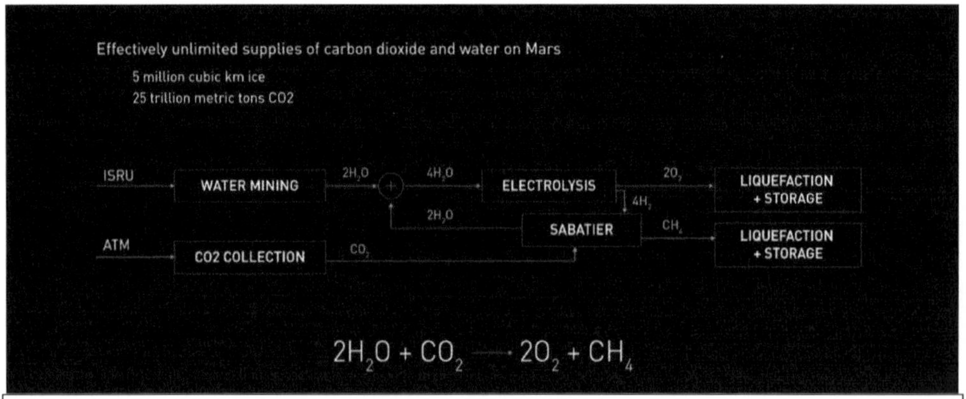

Mars ISRU propellant plant process, credit SpaceX

"And then on Mars, because the atmosphere is carbon dioxide and there's a lot of water or ice in the soil, the carbon dioxide gets you CO2, the water gives you H2O. With that you create CH4 [methane] and O2, which gives you combustion. So it's all sort of nicely worked out. And then one of the key questions is can you get to the surface of Mars and back to Earth on a single stage. The answer is yes, if you reduce the return payload to approximately one-quarter of the outbound payload, which I thought made sense because you are going to want to transport a lot more to Mars than you'd want to transfer from Mars to Earth. For the spacecraft, the heat shield, the life support system, and the legs will have to be very, very light(22)." ~ *Elon Musk*

Entry velocity at Earth should be even higher (>12 km/s), though most of this velocity will be lost through aerobraking at higher altitude, allowing a more moderate (2-3g) propulsive landing at the original launch site.

If they are able to reduce transit time down to three to four months, the Starship spacecraft should, in some cases, be able to complete a round trip using the same planetary conjunction window, resulting in a 6 - 8 month overall flight duration. This should provide sufficient time for the spacecraft to be refurbished, reloaded and refueled, ready to transport more people and cargo to Mars during the next planetary conjunction window (which occurs approximately every twenty six months).

Starship Cargo

Where possible the Mars colony cargo will be 'flat-packed' and stored in the spacecraft's hold. The Starship has over 1000 cubic meters of pressurized volume in the forward section for cargo and/or crew, with additional space available for unpressurised cargo at the rear of the vehicle.

Starship Launch Abort System

The Launch Abort System (LAS) is designed to rapidly separate Starship in the event Super Heavy booster fails during launch. The Starship's Raptor engines would be used to perform this abort maneuver then safely guide it to a propulsive landing. Raptor uses methalox torch igniters to quickly start the engines, allowing Starship to power away from any booster trouble.

"Raptor turbines can spin up extremely fast. We take it easy on the test stand, but that's not indicative of capability(23)." ~ *Elon Musk*

It's probably important to note that Starship will need to fire all its Raptor engines to supply sufficient thrust for this abort maneuver. Fortunately, the sea level and vacuum engines are both capable of operating in space and at ground level (in a pad abort scenario). This is possible because the Raptor vac engines are fitted with dual bells, which allows them to work over the entire range of atmospheric pressures experienced, from takeoff through to orbit. It's possible SpaceX may opt to use Raptor-Max vacuum engines to provide cleaner stage separation, once they become available.

Starship Landing System

Starship lands propulsively on Mars, credit SpaceX

All being well the Starship spacecraft will majestically descend to its designated landing spot, suspended on three (deep throttled) Raptor engines. If, for any reason, one of these axial engines fail, it should be possible to land using the two remaining engines. Unfortunately aborting to orbit wouldn't be possible due to limited propellant, however, having triple redundant axial engines with three peripheral engines available in reserve should almost guarantee a safe landing.

Radiation Mitigation

From past experience, exposure to ionizing radiation (originating from the sun and the larger cosmos) presents a serious problem for Beyond Earth Orbit (BEO) missions. For example, Apollo astronauts have 4-5 times the expected rate of heart disease(24) compared to astronauts who didn't leave Earth orbit. Clinicians certainly try to avoid whole body irradiation (except where absolutely necessary) due to the many deleterious effects (increased risk of cancer and heart disease being chief amongst many). Unfortunately, radiation pervades space, to a greater or lesser degree, hence it might appear a showstopper for long duration missions, such as voyages to Mars. However, SpaceX have a relatively innovative approach to mitigate the effects of ionizing radiation, using improved space materials, reduced exposure time and optimal vehicle orientation.

Apollo 15 Spacecraft in lunar orbit, credit Wikipedia

Traditional spacecraft, like Apollo, were mainly constructed from light aluminum. This material provides adequate protection from solar radiation, which primarily consists of protons, electrons and helium nuclei ejected at high speed from the sun. However, SpaceX intend to construct Starship from stainless steel, a more durable material with improved radiation management properties – vital for long duration missions. This should also provide a better way to handle cosmic radiation (ranging from free electrons up to heavy nuclei travelling at relativistic speeds after being ejected by exploding supernovae), compared to light metal shielding. Normally, if cosmic radiation impacts any low mass materials, like an aluminum hull, it can produce a shotgun-like burst of secondary radiation inside the spacecraft. However, stainless is high mass, hence the atoms are more widely spaced, making it easier for cosmic radiation to pass cleanly through the spacecraft, hence reducing the risk of secondary radiation.

In addition, Starship should maintain optimum orientation during flight, with its heat shield facing the sun. This should allow the double layer of stainless steel and ceramic heat tiles to act as a radiation shield against solar radiation – layers are particularly effective against such radiation. In addition, they will likely maximize the protective effect by having a relatively low angle of incidence, effectively increasing the thickness of the shielding material.

Last but not least, Starship should transit to Mars in 3-6 months, which promises to greatly reduce radiation exposure compared to a more conventional 8-9 month Hohmann transfer.

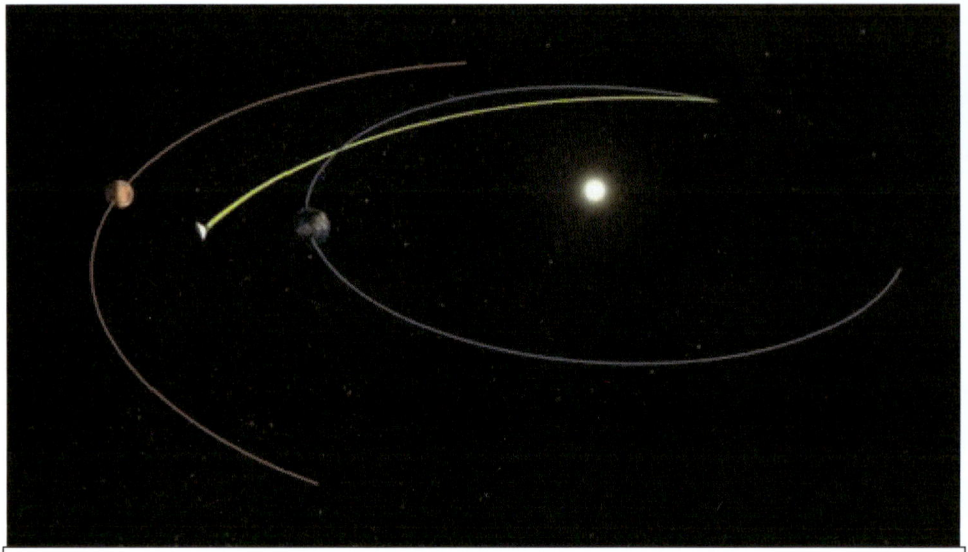

Hohmann trajectory from Earth to Mars, credit openNASA

Overall, these radiation mitigation measures should prove superior to previous techniques and adequately protect passengers from everything including full blown solar storms. For relatively brief periods, increased activity on the sun produces solar radiation levels far in excess of what is tolerable. Radiation from these storms might overcome or even circumvent the inherent protections, in which case passengers will need to take shelter in Starship's central column, which can be fortified against radiation, until the storm subsides.

Generally, the human body is relatively tolerant to low levels of radiation, right down to the DNA level. Radiation only becomes a problem at high intensity, when it can effectively overwhelm the body's self-repair mechanisms. NASA estimate on a six month journey to Mars people might receive on average 300 mSv (millisieverts) of radiation(25), so for a journey of half that time (which SpaceX intend) passengers should receive ~150 mSv – which is comparable to the exposure for a six month stay on the ISS. No doubt the other radiation mitigation measures SpaceX employ will further reduce the expected exposure - as we've already discovered, Starship is far from a standard model spacecraft.

Zero-G Mitigation

Cosmonauts Mikhail Kornienko, Sergey Volkov and astronaut Scott Kelly return after one year mission, credit NASA

Prolonged exposure to zero gravity can also debilitate space travelers because low or no gravity usually results in a reduction in bone density and muscle wastage. These physiological changes have little effect on physical performance while in space, however, a rapid return to full gravity can be hazardous (causes feinting, nausea, loss of balance, increased fragility, low energy etc). A program of resistive exercise can help offset these space adaptations(26), allowing people to rapidly recover, once they return to full gravity.

SpaceX have suggested the Starship entertainment area could be reconfigured for zero-g games and exercise. Overall, the positive effects of exercise and the relatively short flight duration should result in little debilitation on Mars, due to its inherent low gravity (0.376 g) compared to Earth.

STARSHIP OPERATION CONCLUSIONS

1. SpaceX plan to reuse the Super Heavy booster to place the maximum payload into orbit in the shortest time possible. This key advance promises to reduce the launch cost by 10-100 times compared to expendable rockets, making Mars flights operationally practicable and economically feasible.

2. The Starship Launch System provides an elegant engineering solution to the challenges of planetary colonization. Spacecraft reuse after orbital refueling, supersonic retropropulsion and in situ resource utilization are all significant advances in space technology and key enablers for the colonization of Mars.

3. Major obstacles to deep space flight, i.e. prolonged exposure to radiation and microgravity, have effectively been removed through use of new materials, fast transit time and an improved understanding of the benefits derived from resistive exercise.

4. No doubt there's a lot of technical details left to pencil in before Mars colonization flights can proceed, but SpaceX's broad brush strokes are bold and appear to cover most of the canvass.

[1]http://www.universetoday.com/14824/distance-from-earth-to-mars/
[2]https://www.youtube.com/watch?v=tdUX3ypDVwI
[3]https://www.youtube.com/watch?v=zu7WJD8vpAQ
[4]https://www.youtube.com/watch?v=sOpMrVnjYeY
[5]https://twitter.com/elonmusk/status/1149571338748616704
[6]https://www.reddit.com/r/spacex/comments/590wi9/i_am_elon_musk_ask_me_anything_about_becoming_a/d94ub7h/?context=3
[7]https://youtu.be/x2gyo1hbheE?t=702
[8]https://twitter.com/elonmusk/status/1166856662336102401
[9]https://twitter.com/elonmusk/status/1143207768423313408
[10]https://youtu.be/zu7WJD8vpAQ?t=2730
[11]https://twitter.com/elonmusk/status/1131671141624365056
[12]https://twitter.com/elonmusk/status/1143019549492744203
[13]https://twitter.com/elonmusk/status/1143040289587814400
[14]https://twitter.com/elonmusk/status/1143042578973171713
[15]http://www.nasaspaceflight.com/2014/03/spacex-advances-drive-mars-rocket-raptor-power/
[16]https://www.popularmechanics.com/space/moon-mars/a26513651/elon-musk-interview-spacex-mars/
[17] https://twitter.com/elonmusk/status/1154229558989561857
[18]https://twitter.com/elonmusk/status/1143128635525799936
[19]https://twitter.com/elonmusk/status/1183868842843361280
[20]https://www.youtube.com/watch?v=GQueObsIRfI
[21]https://www.reddit.com/r/space/comments/76e79c/i_am_elon_musk_ask_me_anything_about_Starship/dodhupm/?context=3
[22]ISBN: 9780062301239, "Elon Musk: Tesla, SpaceX, and the Quest for a Fantastic Future" by Ashlee Vance
[23]https://twitter.com/elonmusk/status/1171124402726899712
[24]http://arstechnica.com/science/2016/07/apollo-astronauts-dying-of-heart-disease-at-4-5x-the-rate-of-counterparts/
[25]https://www.nasa.gov/pdf/284273main_Radiation_HS_Mod1.pdf
[26]https://www.ncbi.nlm.nih.gov/pubmed/16025596

Chapter 7: Where SpaceX are Building Mars Rockets

Space Shuttle, Falcon 9 (Dragon and Fairing versions), Falcon Heavy, New Glenn (2 and 3 stage versions), Saturn V, SLS, Starship LS, ITS

The challenges associated with building and testing Starship LS presented a whole new set of problems for SpaceX. The stages are so huge they would cost millions to transport from their Hawthorne factory to the port of Los Angeles, then thousands more to ship via the Panama Canal to Cape Canaveral. Once there, the Cape is hardly ideal for intensive development and testing, because the facility is shared with NASA, USAF and various other commercial launch companies. Unfortunately, this means on any working day SpaceX might be required to down tools while these other users perform their own operations, for safety reasons. So what they needed was an entirely new approach for development, where they could build and test Starship LS as intensively as they wanted, preferably somewhere close to the beach.

BOCA CHICA BEACH

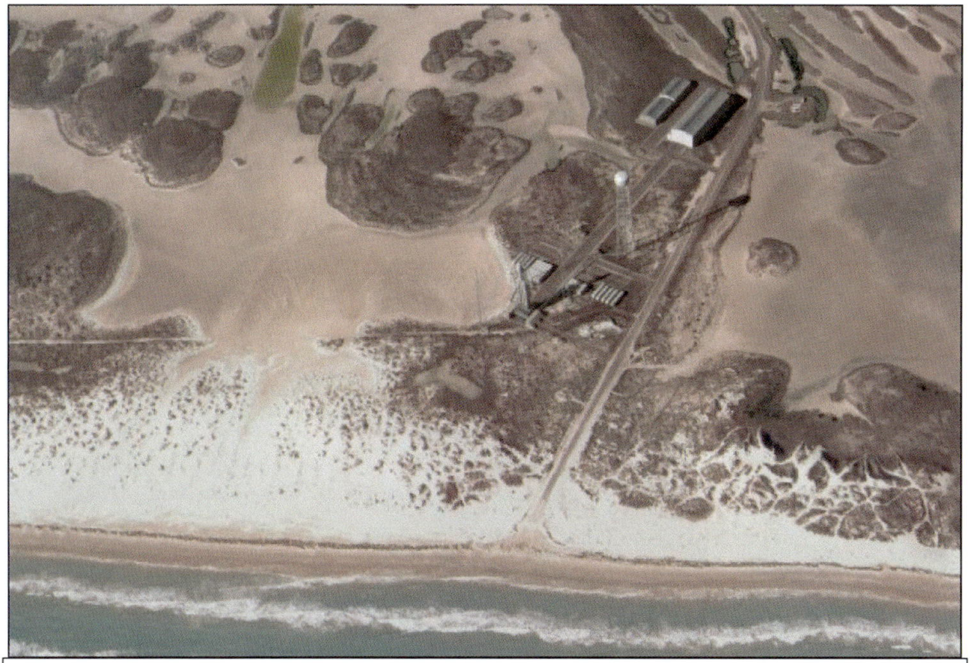

Early depiction of Boca Chica Spaceport, Texas

In August 2014 SpaceX announced they would build their own space launch facility at Boca Chica beach near Brownsville, Texas(1). Originally it was envisioned BC should provide another launch site for Falcon 9 and Falcon Heavy missions, allowing them to increase their annual flight rate.

"The proposed private launch site is needed to provide SpaceX with an exclusive launch facility that would allow the company to accommodate these launches, which have tight launch windows." ~ *FAA Final Environmental Impact Statement: SpaceX Texas Launch Site(2)*

However, this ability to launch whenever they want, would also be a huge boon for Starship LS testing, which will require large numbers of tanking/detanking, static fire and flight operations to be performed at fairly short notice.

Boca Chica becomes breakout site for Starship

"Our south Texas launch site will be dedicated to Starship, because we get enough capacity with two launch complexes at Cape Canaveral and one at Vandenberg to handle all of the Falcon 9 and Falcon Heavy missions(3)." ~ *Elon Musk*

Mars launch from Boca Chica, credit SpaceX

In mid-2018 Elon Musk confirmed Boca Chica would become a dedicated Starship build and test facility. On the face of it, this decision might appear anomalous, because it suggests they intend to construct duplicate facilities (ground support equipment, HIF and launch pad) to those at the Cape. However, BC's location and numerous other advantages make it uniquely qualified as a Starship nursery: -

1. Access – It has good road access (State Highway 4 runs adjacent to the site), suitable for transporting large stages and bulk materials. The nearby Port of Brownsville should be capable of handling these stages (if they need to be transported by sea) and has large liquid methane storage facilities available. Failing that South Bay actually impinges on the launch site if SpaceX wish to build their own port facility.

"It's really a great location. The Cape is at 28.5 degrees latitude. Brownsville is at 26, so you get a little extra boost there [for launches to LEO, GTO or BEO](4)." ~ *Gwynne Shotwell*

2. Location – BC is situated on an east facing coast at the most southerly point of the continental United States. Rockets normally launch over the ocean (for safety reasons) and sites closer to the equator supply a greater assist from the Earth's rotation, allowing more payload to be delivered to LEO or beyond. Considering the experimental nature of Starship work, anything which assists it to reach orbit is of great benefit. Starship is designed to be reusable, so any fuel saved reaching orbit should increase the likelihood of it being recovered and reused.

3. Climate – The climate is relatively moderate at BC, with less average rainfall and storms compared to the Cape(5). SpaceX will need to perform frequent test flights, so BC's easier climate should result in fewer launch scrubs due to adverse weather.

4. Workforce – SpaceX already operate a rocket development and test site in McGregor Texas. As Falcon testing begins to tail-off at McGregor, engineers could transfer to BC as Starship development ramps up. Essentially SpaceX have a readymade workforce in the area raring to work on Starship LS, avoiding the need for an extended and expensive recruitment process.

5. Independence – Freedom to test at Boca Chica is a big advantage. SpaceX own the launch site hence are free to perform construction work and test flights whenever they please. Normally such work would have to be carefully scheduled to avoid conflict with other commercial, civil and military stakeholders at existing sites such as Cape Canaveral. Hence site ownership should allow Starship LS development to proceed at pace, by minimizing any chance of external delays.

6. Environment – Boca Chica has plenty of room to expand, which they will certainly need to accommodate the prodigal Starship LS. Environment effects, such as noise from engine trials and test flights are a concern but any disruption should be minimized due to the sparse local population.

7. Support – Building a Starship launch facility at BC is expensive and SpaceX will have to contribute the lion's share of the money. However, it seems likely they might be awarded some local/state subsidies, similar to the support they already receive to establish the BC launch site.

"About a year ago, SpaceX came to me with their concept of a new, larger, expanded plan for Boca Chica Beach. The concept went well beyond conducting launches, and would require new commitments for construction, investment and jobs to support the new operations(6)." ~ *State Rep. René Oliveira, D-Brownsville*

8. Isolation – Last but not least, if Starship LS tests are performed exclusively at BC this should allow launch operations to proceed uninterrupted at their other sites. SpaceX relies on a steady stream of commercial revenue to finance Starship LS development, so any way to minimize potential disruption could be viewed as essential.

Expanding Boca Chica to accommodate Starship wouldn't be without its difficulties, however. During the 2014 groundbreaking ceremony, the SpaceX communications director John Taylor advised: -

"'Imagine a football field, now imagine that football field thirteen stories tall. That's how much soil is needed to stabilize the foundation. This process is called soil surcharging, and the soil will have to be trucked in, because there's no bedrock, nothing to build on. They dug three hundred feet beneath the shore and hit nothing, just rocky mountain silt built up over millennia(7)."

Going to Mars won't be easy but SpaceX are committed to going those extra yards.

Starship LS Production Facilities

Starship assembly at Boca Chica Texas

SpaceX are currently constructing Starship at two locations: Boca Chica in South Texas(8) and at Cocoa in Florida(9), not far from Cape Canaveral. These stages are quite bulky and hence difficult to transport so they opted to build them as close as possible to the launch sites to reduce risk and speed development.

In an original approach, they decided to construct these vehicles in the open air, similar to a shipbuilding process. First stainless steel plates are welded together to form 9m diameter ring or nosecone sections then each section is stacked and welded, to create either a Super Heavy booster or Starship second stage. This construction technique has proved fast and low cost, keeping development on track for full stack Starship test flights in 2020.

Starhopper (aka Starship prototype) Development

Starhopper test vehicle during construction

In December 2018, SpaceX began to construct a flying testbed, called Starhopper, at their Starship construction yard opposite Boca Chica village. The vehicle's size (39m high and 9m diameter) is similar to Starship's hence possesses comparable flight characteristics. Starhopper was primarily designed for short hop flights, to test take-off and landing. It performed two successful static fire tests in April, to ensure the Raptor engine performs nominally, with the stage chained down to the launch area (comprised of a concrete apron, similar to LZ-1 at the Cape). This was followed by flight tests in July (18m altitude) and August (150m altitude with 100m horizontal divert), to practise low speed maneuvers, all without crew onboard (strictly autopilot control). Rather than risk it further, SpaceX then chose to retire Starhopper and now use it solely as a landmark and occasional engine test stand. Likely this will be the only Starhopper built as SpaceX rapidly proceeded to construct the first two Starships (Mark 1 at Boca Chica and Mark 2 at Cocoa), while Starhopper tests were still being carried out.

Starship Development

"Most likely it [hopper tests] will happen at our Brownsville location...by hopper tests I mean it will go up several miles and come down, the ship is capable of single stage to orbit if we fully load the tanks, so we'll do flights of increasing complexity. We will want to test the heat shield material, fly out, turn back, accelerate back real hard and come in hot to test the heat shield(10)." ~ *Elon Musk*

Prototype Starship during Earth re-entry, credit SpaceX

Following Starhopper trials, they should commence flight tests of the Mark 1 Starship. All being well, they will attempt to perform long suborbital flights over the Gulf of Mexico, ending in a fast powered return to test heat durability under simulated re-entry conditions. Quite possibly SpaceX will finish testing with some long range sub-orbital flights, using Starships ability to reach space without the help of a booster stage (sans payload). Barring mishaps, these flight tests could be complete in less than a year and require only one or two Starship test vehicles, due to its design capability for full and rapid reuse. Basically, they built the primary test vehicle at BC – and one just in case at Cocoa, for use at the Cape. This should give both sites experience of building and operating Starship before they proceed to full stack Starship LS.

Super Heavy Development

During the 2019 update, Elon confirmed Super Heavy will also be built and tested at Boca Chica. This will require a full Starship LS pad to be constructed at BC which will utilize the same ground support equipment used for Starship tests. The pad consists of a raised launch platform and sound suppression system, with a level apron situated nearby to handle stage landings. Short hop tests are to be expected, followed by lofted flights over the Gulf, ending in a high-speed approach and landing, to simulate return from orbital missions.

Booster stage landing next to launch platform, credit SpaceX

In a similar way to Starship development, Super Heavy should also be constructed at Cocoa Florida then moved across the river to the Cape, ready for full stack operation, once flight testing is complete at Boca Chica.

Starship Launch System Development

After Starship and Super Heavy fly separately, SpaceX should then proceed to test the full stack vehicle at their dedicated launch facility at BC. This Starship LS launch complex is designed to be highly robust and quickly refurbished, hence ideally suited to these inherently risky trial operations. For example, the Falcon 9 requires a TEL (Transporter Erector Launcher) mechanism to maneuver the complete vehicle onto the pad and raise it vertical. The TEL is also used to load propellant onto the vehicle and continuously top up the oxygen tanks, to replace any propellant lost to boil-off. This process continues right up to the moment of launch, at which point the TEL receives a fair amount of damage due to its proximity to the exhaust. However, SpaceX have produced an innovative design for a dedicated crane system, which they intend to use for lifting Starship LS stages onto the launch platform, eliminating any need for a TEL. First the Super Heavy booster will be hoisted high onto the platform and secured in the hold down cradle. Then the same crane mechanism will be used to lift Starship and place it on top of Super

Heavy, in a process known as vertical stage integration. Propellant is then pumped into the rocket through lines attached at the base of vehicle. When the booster tanks are full, propellant is then pushed up to Starship's tanks at the top of the vehicle, via the same couplers required for orbital refueling.

After propellant loading is complete, they should be ready to perform a static fire test. For a brief period all Super Heavy engines will be lit and spool up to maximum power, while the vehicle is securely clamped to the platform. Assuming full thrust is achieved, and all systems perform nominally, Starship LS should then be ready for launch. We can only imagine the spectacle that will bring – the first Mars capable vehicle literally panting for lift-off.

BFR prepares to launch, credit David Romax/Gravitation Innovation

When this prodigious launch complex is complete, the construction yard, dedicated roadway and launch mount will stretch for miles, extending all the way from Boca Chica village to the beach.

Starship LS Tests

SpaceX intend a surprising brisk test program for the Starship Launch System. Here's what we know of the test schedule so far: -

Starship LS Test Schedule	
Test Date	**Test Description**
NET* December 2019	Starship Mk.1 high altitude (20km) flight, combined with heat shield proving test
NET* April 2020	First full stack maiden flight
NET* October 2020	First crew flight
NET* July 2021	First lunar landing(11)
NET* July 2022	First crew lunar landing(11)

Whether they achieve this ambitious schedule has yet to be seen, but the progress they managed over one year is astounding. In September 2018 they had little more than a mound of dirt at Boca Chica, now they have a 50m Starship. Only SpaceX...

*NET (No Earlier Than) – the earliest possible date, assuming no major issues in development, construction or obtaining flight clearance.

Floating Spaceport

Once SpaceX have finished the Starship LS launch facility at BC they may move on to construct a seagoing spaceport to handle Starship launch and landing operations. They intend Starship to carry commercial passengers around the world (see Chapter 8, Point to Point Transport), which will likely require floating Starship spaceports to be situated in international waters, adjacent major cities. Hence it is quite possible SpaceX may operate a Starship 'lilypad' from BC in order to develop this point-to-point capability. Likely these seagoing spaceports will each require a small fleet of support vessels, including a dedicated propellant tanker (capable of carrying deep cryo methane and oxygen fuel) and a fast passenger transfer vessel. Floating spaceports might appear to add unnecessary complexity to Starship operations but they are essential to reduce any noise effects on built up areas (Starship operations will be relatively high volume, particularly the sonic boom prior to landing).

Long term future of Boca Chica

Once established, SpaceX will continue to use Boca Chica for Starship operations. They require a large number of operational flights over a relatively short period (e.g. to launch Tanker Starships), so the additional capacity offered by BC should be invaluable to maintaining a high flight cadence. The construction site too will produce Starship LS stages continuously, to increase the number of Mars colony flights, support satellite launch services and build up a fleet of point-to-point passenger vehicles.

"Now over time there would be many spaceships. Ultimately, I think, upwards of 1,000 or more spaceships waiting in orbit. And so the Mars colonial fleet would depart en masse [during each Earth-Mars conjunction window](12)." ~ *Elon Musk*

CONCLUSIONS

1. The fact SpaceX didn't wait for Starhopper to complete testing before moving ahead with Starship construction speaks volumes. It means SpaceX are truly committed to building the first interplanetary spacecraft, for them it can't come soon enough.

2. Their decision to build and test Starship at two different launch sites (Boca Chica and the Cape) seems extravagant, considering this effort is privately funded. However, when the process is complete they will have prepared two parallel sites to begin Starship LS production and launch operations.

3. It's quite possible Starship point-to-point will mainly launch from floating spaceports, to minimize noise pollution. If so, it's likely any prototype spaceport will dock at Boca Chica (possibly Brownsville) to ensure testing continues smoothly at this dedicated development site and avoid disruption to other launch sites.

[1]http://www.mysanantonio.com/news/local/article/Texas-SpaceX-announce-spaceport-deal-near-5667434.php#item-38489
[2]https://www.faa.gov/about/office_org/headquarters_offices/ast/environmental/nepa_docs/review/launch/spacex_texas_launch_site_environmental_impact_statement/media/FEIS_SpaceX_Texas_Launch_Site_Vol_I.pdf
[3]https://gist.github.com/theinternetftw/5ba82bd5f4099934fa0556b9d09c123e
[4]http://aviationweek.com/blog/spacexs-gwynne-shotwell-talks-raptor-falcon-9-crs-2-satellite-internet-and-more
[5]https://docs.google.com/presentation/d/17ol1gtgphVAqYsdVWX7oVF4KlaKEC7bgMeIWeo8JXT0/pub?start=true&loop=true&delayms=30000&slide=id.g11c5c50fed_9_0
[6]http://www.brownsvilleherald.com/premium/spacex-funding-request-may-indicate-broader-scope/article_49f9d9ca-06ff-11e8-9d6c-0bb169167e1b.html
[7]http://www.texasmonthly.com/articles/countdown-to-liftoff/
[8]https://www.youtube.com/watch?v=Gn8hZ3hif6w
[9]https://www.youtube.com/watch?v=uzqeSe3Q_zc&feature=youtu.be
[10]https://youtu.be/yzbFqLOjP4E?t=1513
[11]https://time.com/5628572/elon-musk-moon-landing/?xid=tcoshare
[12]https://youtu.be/H7Uyfqi_TE8?t=1385

Chapter 8: How SpaceX Can Source Mars Colony Technology

View from observation deck approaching Mars, credit SpaceX

It's daunting to imagine the mountain of new technology needed just to travel to Mars (Starship etc) but that seems like a mere foothill compared to the Olympus Mons of original tech required for full blown colonization. But, as they used to say in the sixties space race: "there's a man with a plan and a pocket comb," so let's comb through what we know of the plan so far.

In 2002 Elon Musk committed himself to take humanity to Mars when he founded SpaceX. This allowed a select team of engineers and technicians to accumulate sufficient technical acumen and generate adequate funds to undertake the necessary development through commercial enterprise. Seventeen years on, SpaceX is finally in a position to build the spacecraft needed to reach the surface of Mars, proving the viability of this commercial development model. However, an unimaginable amount of raw technology will be required to establish a Mars colony, which threatens to eclipse all their prior achievements.

"People are mistaken when they think that technology just automatically improves. It does not automatically improve. It only improves if a lot of people work very hard to make it better, and actually it will, I think, by itself degrade(1)." ~ *Elon Musk*

Mars Colony Technology

To get to grips with the problem, let's try to compose a wish list for all the advanced technologies they might conceivably need for long term Mars settlement:-

1. **Surface habitat domes** – to provide an amenable living space, essentially a self-contained and supporting biosphere

2. **Mining equipment** – to excavate service areas (which could double as radiation shelters during solar storms) and extract water from sub-surface reserves

3. **Closed ecosystem** – ability to recycle all essential elements such as air, water, nutrients etc, to ensure nothing is wasted and maximize long term sustainability

4. **Surface transport** – pressurized all terrain vehicles used to survey, construct and maintain settlements

5. **Mass transit system** – to transport goods and personnel from Mars spaceport to habitat area and connect remote settlements

6. **Internet system** – to communicate anywhere on planet with minimum delay and minimal surface infrastructure

7. **High bandwidth communications to red planet** – vital for ecommerce and to support colony operations

8. **Artificial Intelligence** – to operate all the autonomous machinery, communications network and coordinate colony activities

9. **Personal Interface** – to monitor and control all systems in the home, work and public environment

10. **High efficiency solar arrays and energy storage** – essential for life support and normal colony operations

11. **Nuclear power** – to provide redundant power for emergency situations and increase supply as the colony grows

12. **Radiation protection** – to allow prolonged surface operations and habitation

A big list which seems pretty daunting but given what we know, there's more than a glimmer of hope. During the original IAC Guadalajara announcement, Elon Musk said: -

"...the main reason I'm personally accumulating assets is in order to fund this. So I really don't have any other motivation for personally accumulating assets, except to be able to make the biggest contribution I can to making life multiplanetary(2)."

Elon Musk and his five sons from first marriage

This could be interpreted to mean Elon Musk intends to personally finance some, if not all of the development costs for Mars colonization. That certainly sounds ambitious, so let's look at the possible ways this could be achieved using the available resources. At present the majority of Elon's assets consist of shares in the companies he helped to create, in particular the large stakes he holds in SpaceX and Tesla. Overall, his net worth has been estimated at $13.2 billion(3), which is certainly a tidy sum. However, if he sold his stakes in Tesla and SpaceX he would effectively cede control to new owners, who might want to change the way they are run and corporate goals. These companies are like Elon's children so abandoning them to an uncertain fate seems unlikely, given his protective and tenacious character.

It is possible Elon Musk might decide to reduce his holdings, for instance he is due to receive $1.4bn in share options from Tesla(4), which could reasonably be liquidated without prejudicing his stewardship of the company. No doubt this Tesla bonus could provide a much needed influx of capital, however, SpaceX burns through $1bn+ in operating expenses each year, so it probably falls several magnitudes short of the outlay required to move forward Mars infrastructure. For practical reasons then, it seems Elon is unlikely to liquidate all his personal assets to finance space colonization. Perhaps the following quote from Elon Musk's biography provides the best clue for what he intends: -

"The visit to Musk Land [SpaceX, Tesla and SolarCity] started to make a few things clear about how Musk has pulled all this off. While the 'putting Man on Mars' talk can strike some people as loopy, it gave Musk a unique rallying cry for his companies. **It's the sweeping goal that forms a unifying principle over everything he does.** Employees at all three companies are well aware of this and well aware that they're attempting to achieve the impossible day in and day out(5)."
~ Ashlee Vance (Elon Musk biographer)

This suggests Elon plans to harness these external assets (Tesla and SolarCity) to assist with the Mars colonization effort, in a similar way to the SpaceX development model. SpaceX was set up with a strong commercial base but chooses to use its profits to develop the key technologies and facilities required for Mars space transport. Considering the effectiveness of this model, this suggests Elon could use all his companies in concert with SpaceX to provide the necessary technical knowhow and financial muscle required for Mars colonization. There has been no official confirmation yet how these broader assets might be employed but we can certainly speculate, given what we know of SpaceX's existing plans and the capabilities of these associate companies.

Mars Habitats

Mars ecosystem enclosed by a geodesic dome

When the first group of Mars colonists arrive, they intend to set up Mars colony habitats which consist of geodesic domes on the surface with extensive subsurface areas to support heavy operations.

"Initially, [we'll use] glass panes with carbon fiber frames to build geodesic domes on the surface, plus a lot of miner/tunneling droids. With the latter, you can build out a huge amount of pressurized space for industrial operations and leave the glass domes for green living space(6)."
~ *Elon Musk*

It is worth noting the surface habitats he describes are similar in construction to the observation deck of Starship, which is effectively a pressurized dome, inset with geodesic glass panels across one half of the dome.

Opera on Observation Deck of Starship spacecraft, credit SpaceX

Due to these structural similarities, it seems likely they intend to build these domed surface habitats at SpaceX, to reduce cost and closely control the fabrication process. Certainly we know they intend to use Starship as the first habitats on Mars, which should allow them to test the latest geodesic glass panel design in the field before they deploy the much larger surface domes.

Mining Equipment

"I think that would be a good idea [bring a boring machine to the moon or Mars]. You could just make as much room as you want underground and you're protected from radiation and everything. And probably use the materials for building and you need to mine ice and dirt anyway, so why not(7)." ~ *Elon Musk/2019 Starship Update*

It seems likely the automated tunneling equipment needed to excavate the subsurface workshops will be sourced externally. In late 2016 Elon Musk announced he intended to create "The Boring Company," which will make automated tunneling equipment to alleviate traffic congestion here on Earth(8). However, it seems likely this new company will also be tasked with developing the requisite boring equipment needed for Mars. Elon Musk usually has larger goals in mind when he starts any new company and it seems an unlikely coincidence he set up a boring company when there's an emerging need for such tunneling equipment at SpaceX. It's possible they will also use this mining equipment to extract large quantities of water from frozen reserves known to exist below the surface of Mars, to provide the raw material needed for in situ propellant production. This should allow them to thoroughly test this new technology before they proceed to excavate the pressurized subsurface areas needed for long term habitation.

Boring machine in SpaceX car park

Interestingly Steve Davis, director of advanced projects at SpaceX, is also project leader for the Boring Company. Creating the ISRU equipment needed for refueling on Mars would be classed as an advanced project, so the person in charge of developing ISRU at SpaceX is also directly responsible for the boring effort...

Closed Ecosystem

Mars is incredibly distant, which means any resupply mission could take up to two years to arrive. For this reason everything they take to Mars will need to be recycled (e.g. air, water, organic waste etc) to ensure the colony is as self-reliant as possible. SpaceX engineers have reportedly approached a NASA funded University of Arizona group involved in developing bioregenerative life support systems for deep space habitation(9). This sort of work would be most applicable to SpaceX's plans for long duration Mars transport and subsequent colonization.

"We're going to put more engineering effort into having a fully-recyclable system for Starship, because if you have a very long journey it makes sense to have a closed-loop oxygen/CO2 system, a closed loop water system, whereas if you're just going out for several days you don't necessarily need a fully-closed loop system(10)" ~ *Elon Musk/#DearMoon announcement*

The colony would also need to establish their own food production to become self sustaining. It has been reported that Kimbal Musk (Elon Musk's younger brother and SpaceX director) has set up a vertical farming incubator called "Square Roots" to encourage young entrepreneurs(11). This venture uses hydroponics to grow fresh agricultural produce in shipping containers, under LED lighting. The main advantage of this vertical farming technique is it supplies the maximum amount of produce using the minimum amount of resources, i.e. water, power and space.

"...the easy way to do the food would be just to do hydroponics. You essentially have solar power—unfurled solar panels on the ground, feed that to underground hydroponics, either underground or shielded by wires, dirt. So then you can be sure that you don't have to worry about excessive ultraviolet radiation or a solar storm or something like that. Really pretty straightforward. You could just use Earth hydroponics(12)." ~ *Elon Musk/Popular Mechanics*

High density cultivation under LED lighting, credit Square Roots

Surface Transport

Most likely some form of electric vehicle will be required on Mars because the atmosphere lacks sufficient oxygen to use internal combustion engines. Fortunately Elon Musk is CEO of Tesla, who currently produce the world's most advanced electric vehicles, at least going by range, performance and Autopilot capabilities. If called upon, Tesla could likely develop much of the key technology required for Mars capable electric vehicles (as part of their normal commercial operation) then transfer them to SpaceX. Cooperation between these two companies should be relatively straightforward considering the Tesla Design Studio is co-located with the SpaceX Headquarters on their Hawthorne campus. Historically there has been a steady exchange of technology and personnel between these two sister companies, hence sourcing electric vehicle tech for Mars is something SpaceXers are unlikely to lose any sleep over.

Render of Tesla electric pickup (Earth edition)

Intriguingly, Tesla drivers have discovered a secret function on their navigation app which allows them to toggle between Earth roads and Mars terrain(13), which suggests Tesla developers are already thinking about Mars.

Mass Transit

Any mass transit system would need to use pressurized compartments and operate in near vacuum, due to Mars's low atmospheric pressure. The Hyperloop system being developed by SpaceX uses self-propelled pods to transport passengers via a vacuum tube hence it could be readily adapted for this role. So far the work on Hyperloop has been limited to building a test track for university teams and research groups to validate their pod designs, as part of a SpaceX hosted competition. However, Elon Musk confirmed at the Hyperloop pod award ceremony that he's certainly considered using Hyperloop on Mars(14).

"On the application of the Hyperloop... on Mars you basically just need a track. On Earth the air density is quite high but on Mars it's 1% of Earth's atmospheric density so probably you might be able to have a road honestly. You'd go pretty fast. But it would obviously have to be electric because there's no oxygen, so you could have really fast electric cars or trains." ~ *Elon Musk*

Hyperloop passenger vehicle

Under normal circumstances, Hyperloop passengers shouldn't need pressure suits, making the whole transfer process between habitations considerably faster and more convenient than conventional surface transport.

Internet System

SpaceX plan to build Starlink, a vast constellation of Earth satellites at LEO, to provide internet connection from space(15). This might technically place them in competition with many of their customers, who also intend to improve connectivity through launching their own communication satellites. Undoubtedly supplying internet to the world could generate a huge stream of revenue, however, SpaceX have a strategic reason for this bold diversification because they intend to use a similar system to provide high speed communication on Mars(16). Having a dedicated, high bandwidth, reliable coms network could be a literal lifesaver on the distant red planet.

At present most of the work to develop these Starlink satellites is being performed by SpaceX Seattle Office. They are currently testing Ka and Ku-bands to communicate with the ground and lasers to interlink between satellites, which should prove considerably faster than conventional fiber optic networks over long distances. The first prototype satellites were deployed in early 2018(17), with a fully operational Earth constellation expected in the mid 2020's. Vast numbers of Starlink satellites will be required to complete this constellation, around 12,000 total. Each satellite transits the sky so quickly at LEO, essentially a whole line of replacement satellites have to follow close behind to ensure a continuous connection. However, a similar constellation over Mars should require much less satellites, perhaps as few as 1,000. Mars's lower gravity allows satellites to fly considerably slower compared to Earth, so the following line of satellites can be much wider spaced, effectively reducing the overall number required. Hence it should be possible to deploy a Mars constellation to provide basic functionality using only a single Starship (300+ satellites), which could then land at Mars Base Alpha. In other words, if such a constellation were deployed at Mars it would be even more practical than Earth's – you might almost say it was made for Mars.

Interplanetary Communications

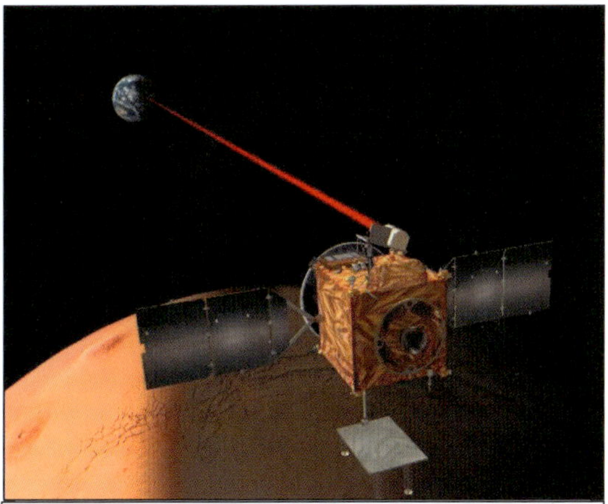
Mars laser communication, credit NASA/JPL

Much of the commerce with Mars will be virtual, e.g. currency transfers, proprietary designs, media products, intellectual property etc, which will need to be transferred electronically. Assuming SpaceX can establish LEO and LMO (Low Mars Orbit) constellations they could use these satellites to connect Earth and Mars via laser communications. This could either be performed by direct connection, i.e. constellation to constellation or by uplinking to more powerful relay satellites parked at Lagrange points. This should dispense with the need for earthbound networks of massive antennae (such as NASA's Deep Space Network) and provide the enormous bandwidth required to support this multiplanetary internet system. In theory anyone on Earth could then communicate direct with Mars via the internet, although a delay of several minutes seems unavoidable due to the intervening distance.

Artificial Intelligence

In 2010 Elon Musk became an early investor in DeepMind Technologies, who produce machine learning programs capable of rivaling humans in specific tasks, such as gaming. Unfortunately, DeepMind was acquired by Google in 2013, effectively closing this possible avenue for AI (Artificial Intelligence) source code.

Elon Musk subsequently co-founded a company called OpenAI which produces open source software to promote AI development, with the long term goal of creating an open source Artificial Intelligence. This highly altruistic approach has allowed them to partner with Microsoft and attract many top AI researchers to the company. In addition, it prompted leading rivals Google and Facebook to open source some of their AI technologies, which further improves OpenAI's prospects of producing their universal Artificial Intelligence.

Given the company's strong financials ($1bn pledged so far) and popular support, it seems likely OpenAI could produce a high machine intelligence or even fully functional AI to assist Mars transit and colony operations.

Personal Interface

In order to optimize effectiveness and safety, almost everything on Mars will be monitored or connected to a computer network (i.e. ubiquitous Internet of Things). Consequently Mars colonists will need to be equipped with a dedicated personal interface device to function in this IoT rich environment. Any delay or lack of computer access would certainly reduce individual effectiveness and possibly hazard their own or other colonist's survival.

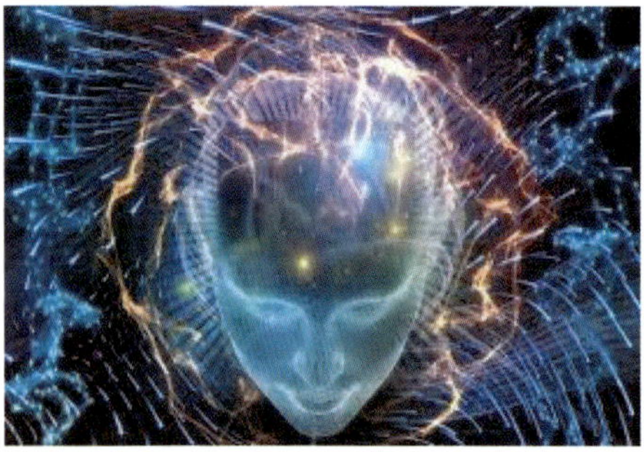

In March 2017, Elon Musk announced he was starting a company called Neuralink(18), to produce an electronic implant which interfaces directly with the brain (called a neural lace). Initially this could be used to help disabled people control prosthetic limbs or receive sensory information. A neural lace could also allow direct access to the internet, without need for external hardware. Effectively this would remove the information bottleneck caused by our slow and cumbersome computer interface systems (e.g. computer keyboards, monitors, touch screens etc) allowing computer access at the speed of thought. This extreme information integration could be essential for survival on Mars, when the difference between life and death literally depends on speed of action and response. Certainly, a neural lace would be a godsend for anyone working in a pressure suit, who might otherwise have to manipulate touch screens and manual controls using thick pressurized gloves.

This technology should eventually allow us to merge with our own personal AI, producing a transorganic symbiotic entity. This could be viewed as the next step in human evolution enabling unimaginable leaps in mental capacity. For example: colonists could be equipped with augmented senses which allow them to see radiation hotspots or the constituents of air. Truly Martians will be a different species, highly equipped and adapted to all the challenges and opportunities Mars has to offer.

Solar Power

In 2006 Lyndon and Peter Rive (following advice from their cousin Elon Musk) founded a solar energy company called SolarCity. The company supplies high efficiency solar power systems to help reduce energy costs for end users, such as residential, commercial and military customers. Tesla acquired SolarCity in 2016, effectively uniting both companies under Elon Musk's control. SolarCity, in partnership with Panasonic, mass produce hybrid solar panels with high energy conversion(19) and mechanical durability(20). Panasonic also cooperate with Tesla to produce 'Powerwall', a high energy density storage device used to store solar energy for domestic and commercial users.

SolarCity house with hybrid solar panels

As you might expect, a high efficiency solar array coupled with a high energy density storage system would seem the perfect fit for Mars colony operations. On average the sunlight on Mars is around half as strong as on Earth, hence high efficiency solar power conversion is a must. Significant energy will also be required at night and during dust storms, so a Powerwall type system would be essential to make solar power practical on Mars. Quite possibly they intend to use solar power exclusively while the colony is being established, backed by high capacity energy storage. As a contingency they could use fuel cells to provide emergency power but that would consume the propellant normally reserved for the return journey to Earth...truly a last resort.

Nuclear Power

Transatomic molten salt reactor

There are no known hydrocarbon reserves on Mars and precious little oxygen, which removes any possibility of conventional power generation. Solar arrays could provide considerable power but the frequent dust storms on Mars (which can last weeks and cloak the entire planet) would seriously affect the efficiency of these arrays. Hence in the long run there seems little alternative to using nuclear energy, particularly as the colony's power consumption grows. On Earth our ecosystem provides all the complex compounds we require to make fuel, plastics and building materials. However, Mars has no appreciable ecosystem; hence these materials have to be synthesized from basic elements and compounds, which will require a magnitude more energy. Similarly providing suitable life support for colonists (breathable air, purified water, adequate heating) will require much more energy because any closed ecosystem will largely be driven by artificial power. The colony's overall power consumption will likely be in the mega-watt range, which places it firmly in nuclear generator territory.

"To get one ship back, you need about eight football fields worth of solar cells on Mars... It's much better to use nuclear fission reactor... We're working with NASA on that, and hopefully they'll get funding to develop that. They've got a program called Kilopower going that's like, ten thousand watts, a 10 kilowatt reactor. We need a megawatt, but you know, you need to start somewhere(21)." ~ *Tom Mueller, former CTO of Propulsion Development at SpaceX*

Radiation Protection

Mars lacks a magnetosphere, i.e. a magnetic bubble which surrounds the planet, similar to the one which protects our Earth. This strong magnetosphere is the first line of defense against solar and cosmic radiation, helping to deflect it away from the planet or channeling it towards the less inhabited poles where it is mostly absorbed by the atmosphere (resulting in aurora phenomena). However, the residual magnetic effects on Mars are relatively weak and offer much less protection from ambient space radiation. Hence surface habitations and operations on Mars will likely require some additional radiation protection, to reduce radiation levels closer to those enjoyed by Earth.

Elon Musk suggested the best way to provide surface radiation protection is through deploying "localized electro-magnetic field generators(22)," which would effectively surround the whole colony with a mini-magnetosphere. This would largely deflect Incoming radiation away from the magnetically shielded area, allowing it to fall harmlessly beyond the outer bounds of the community.

So far, no additional information has been provided about how such mini-magnetospheres could be created, although it's reasonable to speculate they might involve some kind of super-conducting coil technology, to minimise weight for transport. This technology could be relatively compact, perhaps even small enough to be integrated into the design of Starship. This should allow them to use the same equipment to protect colonists in transit and on the surface, effectively removing radiation as an impediment to long term space colonisation.

Reality Check

Assuming SpaceX manage to achieve their ambitious Mars colony, it will no doubt consume new technology like popcorn. Elon Musk loves advanced tech, so it might be pure coincidence he helped to establish all these high-tech startups with products which could be readily adapted for Mars. Interestingly, Elon Musk has for many years stated he would like to build an electric SST (Super Sonic Transport) aircraft but so far has chosen not to commit to this new venture. Unfortunately, any

SST development would effectively compete with current plans to use Starship for long distance transport and Hyperloop across country. In addition, SST development is somewhat out on its own spiral arm and likely to produce few technologies which could be readily shared with other projects. For example, SST would be impractical on Mars because the thin atmosphere provides insufficient lift for a heavy lift aircraft. In the final analysis SST technology doesn't support other efforts, which probably explains why it hasn't gone forward.

This suggests Elon has certain basic criteria when setting up new companies, as follows: -

1) will it provide significant benefit to society
2) is it commercially viable
3) does its technology support other endeavors

Let's use a table to see if there's any pattern for how these startups are created.

Company Criteria	Space X	Tesla	Solar City	Open AI	Neura Link	Boring Co.	SST
Significant Benefit	✓	✓	✓	✓	✓	✓	✓
Commercially Viable	✓	✓	✓	X	✓	✓	✓
Technology Support	✓	✓	✓	✓	✓	✓	X
Startup Created	✓	✓	✓	✓	✓	✓	X

OK, here's what we can deduce from this table. The OpenAI column seems to indicate if the technology produced by the startup is of significant benefit, and can be used to support other endeavors, commercial viability is not a consideration. Perhaps more significant still, judging by the SST column, how well the technology plays with others seems to be the deciding factor for whether or not these startups are initially created.

It might appear the ARLM ethos used at SpaceX carries over to its sister companies (i.e. All Roads Lead to Mars) but in reality each startup is designed to synergize with its peers, while pursuing its own goals. Perhaps from a SpaceX perspective all roads lead to Mars but these roads also lead back to Earth, likely resulting in great benefit from Musk's more down to Earth endeavors.

"After 5 years of just trying to convince other people to do tunnels, I was like OK I guess I'll do tunnels... You have to have a 3D transportation network to match 3D buildings(23)." ~ *Elon Musk*

Practical Technology Transfer

While the Musk family of companies appear almost perfect for developing all the relevant technologies for Mars, there are some difficulties inherent in transferring these technologies to SpaceX. If, for example, Elon Musk announced the next model produced by Tesla was aimed specifically at the Mars market...Tesla shareholders would likely view this as an unnecessary diversification and clear conflict of interest for their CEO, who is also CEO at SpaceX.

In practice, this might not be a significant problem because of the limited amount of hardware required for initial Mars landings. Quite reasonably, Tesla could sell small batches of components or complete vehicles to SpaceX at near market rates, without ruffling investor feathers. SpaceX could then assemble or adapt the products they need, to their own specification, ensuring they were fully optimized for Mars conditions.

Tesla powered electric locomotive, credit The Boring Company

The Boring Company have already demonstrated how this technology transfer might work in practice. During 2017/18 they purchased a reasonable batch of automotive batteries and motors from Tesla, to power their electric locomotives, used to haul excavated dirt through miles of underground tunnels.

"Elon Musk is a co-founder and significant stockholder of The Boring Company, with which we [Tesla] have entered into an agreement to sell certain vehicle motor and battery pack components. We expect to invoice approximately $400,000 in total for parts sold and to be sold to The Boring Company during 2017 and 2018(24)." ~ statement to Tesla shareholders

CONCLUSIONS

1. SpaceX haven't officially acknowledged how they intend to source the technologies for Mars colonization but have an array of suitable commercial assets they can call on, thanks to the foresight of their enterprising founder Elon Musk. He has created or inspired a consortium of companies and concerns since 2002 which seem capable of delivering all the major technologies required for Mars colonization. As you might expect, he's leaving nothing to chance.

"The synergies are: we [SpaceX] do use Tesla batteries for our technology. We gave Tesla our enterprise information system, we call it Warp Drive... I think the first cars on Mars will be Tesla's.... I think we will be boring tunnels on Mars, I think that's where you gonna live(25)..." ~ Gwynne Shotwell/IE Technology and Innovation Club, Madrid

2. Likely Elon Musk has already achieved a critical mass of companies to attempt space colonization – although it's quite possible other specialty concerns might be added as Mars approaches. Perhaps the way these companies cooperate, while operating in a harsh business environment, might be a glimpse into the future of commercial enterprise in space.

[1]http://uk.businessinsider.com/brilliant-career-advice-from-elon-musk-2017-5?r=US&IR=T
[2]https://youtu.be/H7Uyfqi_TE8?t=2759
[3]http://www.investopedia.com/university/elon-musk-biography/elon-musk-net-worth.asp
[4]https://www.bloomberg.com/news/articles/2017-04-20/elon-musk-nears-1-4-billion-windfall-as-tesla-hits-milestones
[5]ISBN: 9780062301239, "Elon Musk: Tesla, SpaceX, and the Quest for a Fantastic Future" by Ashlee Vance
[6]https://www.reddit.com/r/spacex/comments/590wi9/i_am_elon_musk_ask_me_anything_about_becoming_a/d94t2bv/?context=3
[7]https://youtu.be/sOpMrVnjYeY?t=4906
[8]http://www.recode.net/2016/12/17/13993738/elon-musk-tunnels-dig-traffic-boring-company-twitter-infrastructure-trump
[9]https://www.reddit.com/r/spacex/comments/54zzwe/spacex_is_likely_in_the_very_initial_stages_of/
[10]https://youtu.be/zu7WJD8vpAQ?t=4968
[11]http://uk.businessinsider.com/kimbal-musk-vertical-farms-shipping-containers-2016-8?r=US&IR=T
[12]https://www.popularmechanics.com/space/moon-mars/a26513651/elon-musk-interview-spacex-mars/
[13]http://www.teslarati.com/elon-musk-apparently-baked-mars-based-tesla-easter-egg/
[14]http://uk.businessinsider.com/elon-musk-talks-hyperloop-on-mars-2016-2?r=US&IR=T
[15]http://arstechnica.com/information-technology/2016/11/spacex-plans-worldwide-satellite-internet-with-low-latency-gigabit-speed/)
[16]https://youtu.be/AHeZHyOnsm4?t=432
[17]https://newatlas.com/spacex-starlink-internet-fairing-recovery/53540/
[18]http://uk.businessinsider.com/elon-musk-neuralink-connect-brains-computer-neural-lace-207-3
[19]http://buffalonews.com/2016/11/03/solarcity-plant/
[20]https://www.youtube.com/watch?v=MQb_aTjZ4vA
[21]https://www.reddit.com/r/spacex/comments/6b043z/tom_mueller_interview_speech_skype_call_02_may/dhiygzm/
[22]https://twitter.com/elonmusk/status/1022217710040543232?lang=en
[23]https://youtu.be/e59b8rtAV6Q
[24]https://electrek.co/2018/04/27/tesla-sell-electric-motors-batteries-elon-musk-boring-company/
[25]https://youtu.be/qWPaopcU_hE?t=2438

Chapter 9: SpaceX Run-up to Mars

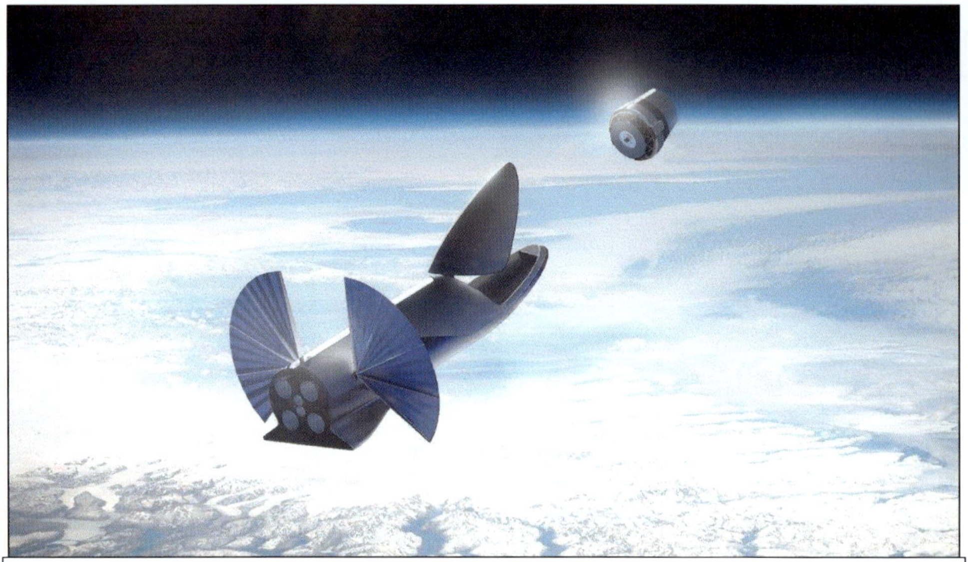

Starship 'Chomper' spacecraft delivers an 8 meter telescope to orbit

At IAC 2017 in Adelaide, Elon Musk revealed SpaceX plans to use Starship in a surprisingly broad range of applications. While this might appear a diversion away from their primary goal of Mars, such work should certainly help to prove the system in a number of innovative and challenging roles. Fortunately, these new roles envisioned for Starship have a sound commercial base, hence should be self-funding and increase revenue available for Mars colonization in the long run. Here's some of the applications SpaceX foresee for Starship in the near future: -

Point-to-Point Transport (aka Earth-2-Earth)

The Starship vehicle is so capable it could be used to fly commercial passengers up to 10,000 km(1), anywhere on Earth. Typically passengers could fly to most destinations in 15-20 minutes, all for the price of an economy airfare. These passenger Starships would be modified for high seating capacity, through multiple decks and typically carry around 1,000(2). Unfortunately these ballistic travelers would have to remain in their seats (due to high acceleration/low gravity issues) and in-flight refreshments could be relatively sparse due to the extraordinarily short flight!

"Probably needs a restraint mechanism like Disney's Space Mountain roller coaster. Would feel similar to Space Mountain in a lot of ways, but you'd exit on another continent(3)." ~ *Elon Musk*

TIME COMPARISONS TO MAJOR CITIES

ROUTE	DISTANCE	COMMERCIAL AIRLINE	TIME VIA BFR
LOS ANGELES TO NEW YORK	3,983km	5 hours, 25 min	25 min
BANGKOK TO DUBAI	4,909km	6 hours, 25 min	27 min
TOKYO TO SINGAPORE	5,350km	7 hours, 10 min	28 min
LONDON TO NEW YORK	5,555km	7 hours, 55 min	29 min
NEW YORK TO PARIS	5,849km	7 hours, 20 min	30 min
SYDNEY TO SINGAPORE	6,288km	8 hours, 20 min	31 min
LOS ANGELES TO LONDON	8,781km	10 hours, 30 min	32 min
LONDON TO HONG KONG	9,648km	11 hours, 50 min	34 min

It is suggested these sub-orbital vehicles could launch and land at floating spaceports, moored offshore from major coastal cities. Passenger and cargo transfer to these spaceports would be arranged via boat and no doubt SpaceX will fast track passenger registration and boarding procedures, to further reduce overall journey time. To be fair this new mode of transport might present a few problems, which are probably worth exploring.

Export Regulations – Normally any attempt to send advanced intercontinental ballistic technology to other countries would break a variety of US export regulations (ITAR, EAR etc). However, if these floating spaceports were registered US vessels, they would effectively be classed as US territory, making them legal destinations for export purposes. Some kind of private security could be used to protect landed rockets and ensure site security. Host countries would likely view such facilities as highly prestigious and go to lengths to preserve their integrity, as they do with international airports.

National Security – The arrival of an intercontinental ballistic rocket at any large city would seem a showstopper. In the past these vehicles (i.e. ICBMs) were designed to deliver nuclear ordinance instead of passengers, so Starship's arrival might be viewed with trepidation in some quarters. However, anyone wishing to use Starship to deliver a nuclear payload would need to hack the flight computers to ensure the vehicle is redirected to the intended target then smuggle the device past site security. Last but not least, a limited nuclear strike would be counterproductive because it allows the targeted country an opportunity to respond in full...

"We will land [Starship] on our own platform, it's out at sea. Largely because first of all most cities probably won't want that incomer hovering over their billion/trillion dollar buildings/high-rises. Its also very loud, from an acoustic perspective, the sonic booms created by coming back. So we will have to be out at sea, So we will maintain control, we'll probably will be in international waters as well(4)." ~ *Gwynne Shotwell*

Starship lands at Shanghai, credit SpaceX

Commercial SST Viability – From past experience, Supersonic Transport (SST) has been seen to be largely impractical, principally because of issues relating to cost, reliability and environment factors. In Starship's case operating costs should be relatively low because SpaceX are experts at building inexpensive launch vehicles, which incorporate a high degree of reusability. These vehicles are designed to fly to Mars and back without overhaul, so suborbital hops should be relatively straightforward. In addition, Starship uses methane propellant which is very cheap, literally a fraction of the cost of aviation fuel.

SST can cause problems when used overland because the supersonic shockwave could disrupt buildings and cause noise pollution. However, Starship would mostly fly through airless space hence generate no supersonic wake or noise in transit. A sonic boom would announce its arrival but this should occur several miles offshore and over in an instant, compared to conventional aircraft noise which can be quite persistent.

There were accidents associated with SST, namely the Concorde crash at Charles de Gaulle Airport in July 2000 and the TU-144S breakup at the Paris Air Show in June 1973. Investigators concluded the Concorde crash was caused by a piece of debris on the runway which effectively ruptured a fuel line, causing loss of engine power on takeoff. Debris removal shouldn't be a problem at Starship spaceports because the dual launch/landing pad is tiny compared to conventional runways and relatively easy to inspect after each flight. The Tu-144S design seems to have been borrowed from Concorde, however, the intelligence operation used to gain critical design information appears to have been flawed(5). Starship is an enormously advanced vehicle hence SpaceX will have to build and test it from the ground up, therefore it has no peer to distract the design process.

Physiological Problems – Starship passengers will experience some unusual conditions for commercial passengers, namely: moderate g-force during take-off, typically 2-3g(6), followed by 15-20 minutes of low-g, concluding with another brief period of moderate g-force on landing. No doubt SpaceX will ensure passengers are relatively fit, with no serious medical conditions and possibly offer some kind of physiological testing (human centrifuge, parabolic flights etc) to minimize any chance of adverse reactions.

"We are designing it so normal people can fly in it. We'll take care regarding the g-limit, but the experience will undoubtedly be sportier than an airplane(7)."
~ Gwynne Shotwell

Starship will fly at around Mach 20 during the parabolic section of the flight, which is less than orbital velocity, hence passengers will only experience reduced gravity. This should mean passengers largely avoid the negative effects from microgravity, which are principally disorientation and nausea.

POINT TO POINT CONCLUSIONS

1. Point-to-point services will allow Starships to be used continuously (dozens of flights/day possible due to short flight time) and possibly generate significant income to support the Mars effort. Essentially Mars passenger flights might take decades to become commercially viable but point-to-point services should provide good return on investment, due to its comparatively low price, speed and prestige (a great way to gain your 'space wings').

2. Operating a Starship passenger service should provide valuable flight experience and allow the design to rapidly improve, potentially making Mars flights more reliable.

3. SpaceX seems more than capable of handling Starship design and development, judging by their experience with Falcon 9 and Falcon Heavy. Now short hop test flights have begun, we could reasonably expect to see point-to-point commence in the 2020s(8).

Starship delivers to the ISS, credit SpaceX

Starship LS could launch any prospective payload into any Earth orbit and inclination, even direct to GEO with orbital refueling. SpaceX is perhaps the only company which could possibly supply a sufficiently large payload to exceed Starship's launch capacity. For example, they intend to launch 4,425 satellites into LEO to construct their internet constellation, which on paper could be accomplished with only 12 flights of Starship.

Given its low operating cost (estimated at less than $10m per launch) and rapid reusability, all SpaceX launches should rapidly transition to Starship LS as soon as it goes operational. The only problem might be oversupply, for instance Starship could potentially carry a hundred astronauts at a time to the ISS or 100 mt of pressurized cargo, far more than NASA could reasonably require. However, this is unlikely to be a problem because Starship could easily launch to the ISS with a reduced payload or possibly go on a round trip, delivering additional passengers and cargo to commercial space stations on the same vehicle (NASA astronauts will no doubt call shotgun).

Moon Base

Starship lands at Moon Base Alpha, credit SpaceX

"Based on the calculations we've done, we can actually do lunar surface missions, with no propellant production on the surface of the moon. So if we do a high elliptic parking orbit for the ship, and retank in high elliptic orbit, we can go all the way to the moon, and back, with no local propellant production on the moon. That would enable the creation of Moon Base Alpha or some sort of lunar base(9)." ~ *Elon Musk at IAC 2017*

The prospect of SpaceX building a base on the moon might appear an unnecessary distraction considering their primary goal is to colonize Mars. However, this makes more sense if you consider the political context.

"We will return American astronauts to the moon, not only to leave behind footprints and flags, but to build the foundation we need to send Americans to Mars and beyond. The moon will be a stepping-stone, a training ground, a venue to strengthen our commercial and international partnerships as we refocus America's space program toward human space exploration(10)."
~ *Vice President Mike Pence, Chairman of the National Space Council*

From all accounts(10)(11), landing on the moon will become a priority for the White House. For political and practical reasons this project will require a significant amount of progress before the end of President Trump's first term, and if re-elected, one or more successful Moon landings during his second. Harnessing NewSpace companies might be the only practical means to achieve this ambitious schedule and SpaceX are certainly well placed, considering their previous success and in-depth preparation. Certainly, Starship LS has no rival for lift capability and the first test flights should occur in 2020, around election time... SpaceX seem ideally placed to lead the charge to the moon which could be seen as a necessary diversification due to numerous benefits: -

1. Building a lunar base would grant practical experience with creating off world settlements, which should prove invaluable for Mars.

2. Some craters at the lunar poles contain millions of tons of frozen water, carbon dioxide and methane(12). These in situ resources could be used to produce methalox propellant on the moon (like SpaceX intends on Mars) making it an ideal proving ground for this technology.

3. The moon has an exceptionally low gravity (around a sixth of Earth's) hence a Starship tanker launched from the moon could carry much more propellant to Low Earth Orbit than a full stack Starship launched from Earth. For example, it should require only one lunar tanker flight to refuel a Starship in LEO, compared to four or more Starship tanker flights originating from Earth.

4. A moon base should be highly desirable to NASA, who would likely support any SpaceX development. The cost of any supply flights would also be paid by NASA or possibly split with ESA, who have already proposed building a multinational "Moon Village." This should support SpaceX efforts to develop Mars technology and provide a profitable destination for Starship operation.

5. Mars flights are only possible every 26 months, while flights to the moon could be operated daily. Hence using Starship for Moon flights first would make maximum use of this invaluable launch system, while they wait for the next Mars launch window.

"For sure moon [landings] first, as it's only 3 days away & you don't need interplanetary orbital synchronization(13)." ~ *Elon Musk*

Considering the political and commercial sensitivity concerning building offworld bases, it's hardly surprising SpaceX haven't made public any of their plans – so far. However, Elon appears optimistic about the prospect, suggesting a Moon base could be built by 2025(14), followed by a Mars base in 2028(15).

CONCLUSION

SpaceX have come up with a mouth-watering confection of possible applications for Starship, many of which demonstrate new or enhanced capabilities. However, most of these applications would depend on government support or approval, which might prove difficult to procure. While point-to-point transport and Moon colonization appear optimistic, their likelihood will no doubt improve as Starship operations mature.

[1]https://twitter.com/elonmusk/status/1134025184942313473?s=19
[2]https://twitter.com/elonmusk/status/1144004310503530496
[3]https://twitter.com/elonmusk/status/1144005431020392450
[4]https://youtu.be/qWPaopcU_hE?t=1960
[5]https://www.quora.com/What-are-some-of-the-mind-blowing-operations-of-MI6
[6]https://twitter.com/elonmusk/status/913775198423408640
[7]https://www.reddit.com/r/spacex/comments/75ufq9/interesting_items_from_gwynne_shotwells_talk_at/do94zn3/
[8]https://youtu.be/Dar8P3r7GYA?t=1005
[9]https://www.reddit.com/r/spacex/comments/73cw1u/my_notestranscript_elons_iac_2017_talk_parts_1_2/
[10]http://spacenews.com/national-space-council-calls-for-human-return-to-the-moon/
[11]https://arstechnica.com/science/2017/10/its-official-trump-administration-turns-nasa-back-toward-the-moon/
[12]https://science.nasa.gov/science-news/science-at-nasa/2010/21oct_lcross2
[13]https://twitter.com/elonmusk/status/1143007430898528256
[14]https://twitter.com/elonmusk/status/1055350177941225473
[15]https://twitter.com/elonmusk/status/1043253619485622272

Chapter 10: Where SpaceX Will Launch to Mars

At present SpaceX intend to operate a two tier system for Starship LS development and flight operations. Primary development is carried out at their facility at Boca Chica, and once complete, full stack launch operations should commence from the Cape.

CAPE CANAVERAL SPACEPORT

Mars rocket prepares to launch from LC-39A, Cape Canaveral, Florida, credit SpaceX

SpaceX have agreed a twenty year lease with NASA for the historic Launch Complex 39A, previously used to launch numerous Space Shuttle and Apollo missions, including the first lunar landing. The 39A pad is currently employed to launch Falcon 9 and Falcon Heavy missions, including crew demonstration flights in 2019/20. SpaceX have already built substantial support facilities, most notably the 100m long HIF (Horizontal Integration Facility) building and a wide gauge inclined elevator railway, to connect the HIF and launch pad.

SpaceX propose to use LC-39A at Cape Canaveral as their primarily Starship launch site, which seems a logical and pragmatic choice. It offers prebuilt ground support equipment, a team of experienced staff and provides an opportunity for NASA to lend their direct support and guidance.

"In the future, there may be a NASA contract, there may not be, I don't know. If there is that's a good thing, if there's not probably not a good thing, because there's larger issues than space here, are we humans gonna become a multiplanetary species or not(1)?" ~ *Elon Musk*

A great deal more work will be required to convert LC-39A for Starship LS, e.g. converting the pad to deep cryo methane fueling, vertical stage integration and facilitate precision landings.

LC-39A modifications

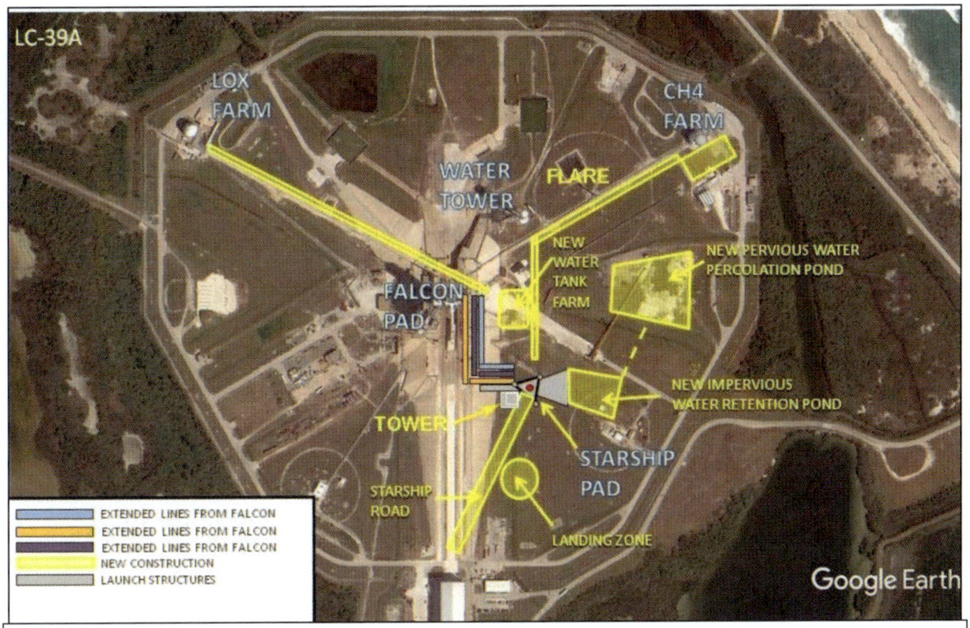

Overview of new Starship LS Pad at LC-39A

Starship LS will require a new purpose built launch pad to be constructed near the existing LC-39A launch mount. The Saturn V, for comparison, generated 3,579 mt (metric tons) of thrust and weighed 3,059 mt, whereas the Starship LS will produce 7,400 mt of thrust and weigh 5,000 mt at launch. SpaceX plan to build an elevated launch platform (over a flame diverter) to support a colossal launch cradle. In addition, a landing zone (similar to LZ-1) will be located nearby to receive returning stages. At present they intend to land both Starship and Super Heavy stages downrange on a specially adapted ASDS, until they perfect the landing technique.

Starship LS Transport

Starship LS stages are too large to be moved far by road, so SpaceX intend to use barges to transport them from their production facility at Cocoa to Cape Canaveral. These barges will be offloaded at the Turn Basin, which was originally designed to receive Saturn V stages, adjacent to the VAB (Vehicle Assembly Building) then transported via truck to the HIF at LC-39A.

Starship LS Tests

After the Starship LS stages have been processed at the HIF, they will need to perform a series of hot fire tests. Likely these stages will be transported in turn to the adjacent launch pad via truck and hoisted onto the pad using a mobile crane. Then the engines will undergo hot fire tests of increasing duration, while being clamped down securely to the launch/landing cradle. All being well, these stages will then be vertically integrated on the pad using the same crane equipment.

"The rocket would be integrated vertically on the pad at LC-39A using a mobile crane. This would involve the booster being mated to the launch mount followed by Starship being mated to the booster. Initial flights would use a temporary or mobile crane, with a permanent crane tower constructed later. The height of the permanent crane tower would be approx. 120 to 180 m(2)."
~ *Environmental Assessment for SpaceX Starship and Super Heavy Launch Vehicle at KSC*

Launch/Landing Pad

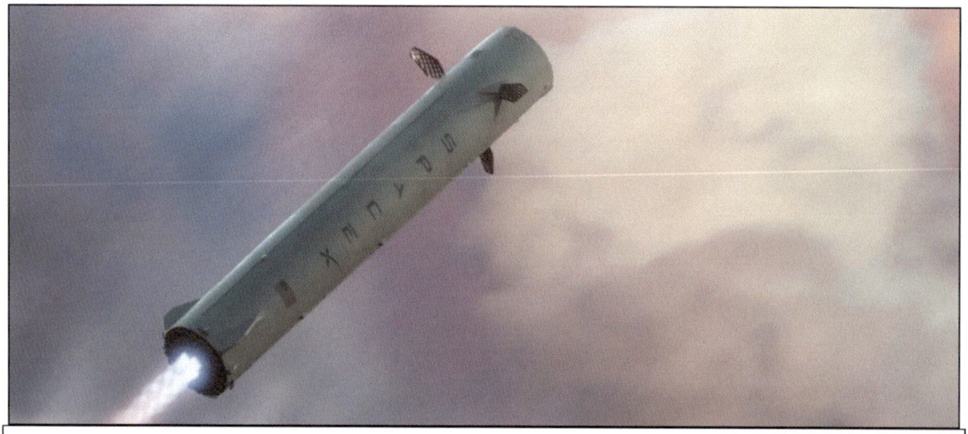

Booster stage approaching landing, credit SpaceX

While landing Starship LS stages on an ASDS is practical in the short term, SpaceX have a stretch goal for recovering these stages at the Cape. Once ready, they intend to land them back at LC-39A, at the adjacent landing zone. This should remove any delays from transshipment and place these stages in the optimum location to be refurbished, ready for their next flight. Minimizing turnaround time should allow more frequent Starship LS launches; particularly useful for refueling flights which could rapidly reuse the same pad and booster.

However, this goal is ambitious, even for SpaceX, and requires some serious challenges to be overcome: -

- Landing accuracy must be assured, to ensure returning stages don't damage adjacent ground support equipment

- Starship LS stages should require little to no refurbishment after each flight to justify time saved landing them at the launch site

- Authorities (primarily NASA and FAA) must give their approval for this untried and possibly risky maneuver – taking into account the proximity of the crew launch pad and other vital Cape facilities

All told SpaceX engineers have a lot on their plate to attune LC-39A to this new reality for space access.

ADDITIONAL LAUNCH SITES

At the south Summit 2015, Gwynne Shotwell (SpaceX COO) stated they will need: "a lot of launch sites flying a lot of rockets(3)," during the tight time-frame for Mars flights. It could require up to five Starship tanker deliveries to fully refuel the Starship spacecraft in orbit, with such refueling flights executed relatively quickly to avoid propellant boil off. Possessing more sites at different locations should certainly assist this process, because they could operate in parallel and offer staggered launch windows to rendezvous with orbiting Starships. At present Boca Chica is the only other confirmed launch site for Starship – although details are scant.

"Starship will do orbital launches from Boca Chica (near Brownsville) & Cape(4)" ~ *Elon Musk*

While additional launch sites have yet to be selected to support these auxiliary missions, it might be possible to launch from multiple offshore spaceports, reducing the need for land-based facilities. As previously discussed, SpaceX are planning to transport commercial passengers all around the world, using a network of offshore spaceports, similar to their existing ASDS only larger. Once this network is established, it might be possible to retask these spaceports to support Mars colony flights, during each Earth-Mars conjunction. Typically, they could launch empty Starships and propellant tankers to LEO in preparation for the next Earth-Mars window.

"If we're loading [Starship] spacecraft into orbit with crew compartments a year in advance, you would[n't] want people on board then. So we would probably launch a mission that just transfers people. So it would wait until the spacecraft is fully fueled, then launch another one that transfers people from tanks empty to one with tanks full(5)." ~ *Elon Musk*

SpaceX will require plenty of launch sites and tankers to perform the necessary large numbers of Starship flights during these relatively short Earth-Mars launch windows and will probably make full use of their available resources. Eventually Elon Musk wants to send a million people to Mars, so a network of launch sites spanning the globe should certainly be useful. No doubt Mars colonists would prefer to embark relatively close to their home nation, particularly if their prior work was 'security sensitive' e.g. rocket scientists, nuclear engineers or submarine specialists (see Chapter 14: Who SpaceX Will Send to Mars).

CONCLUSIONS

1. At IAC 2016, Elon Musk proposed ITS should launch from Cape Canaveral – which unfortunately received a subdued response from NASA. Then at IAC 2017 he presented a new low-cost architecture for BFR, which SpaceX could fund themself, through normal commercial revenue. Given all the benefits Boca Chica has to offer, it seems likely comprehensive testing will be carried out there, then proceed to crew flights from the Cape.

2. Once demand for Starship LS increases (e.g. for tanker flights), Boca Chica and Cape Canaveral will likely work in parallel, with crew flights operated from the Cape.

3. SpaceX have some high hurdles to clear before they can perform stage landings at LC-39A.

[1]http://toaster.cc/2016/10/04/IAC_Press-Conf-Transcript/
[2]https://netspublic.grc.nasa.gov/main/20190801_Final_DRAFT_EA_SpaceX_Starship.pdf
[3]https://youtu.be/omBF1P2VhRI?t=461
[4]https://twitter.com/elonmusk/status/1143335443666280448
[5]http://diyhpl.us/wiki/transcripts/spacex/elon-musk-making-humans-a-multiplanetary-species/

Chapter 11: Where SpaceX Will Land on Mars

The first Mars landing site will likely be located at median latitude i.e. somewhere close to the Mars equator. Equatorial regions have higher average temperatures, more intense sunlight (to optimize solar power collection) and should require less propellant for rocket landings and launches (because of an assist from the planet's rotational speed).

Ideally any prospective sites should also possess water stored in the immediate subsurface, preferably a meter or less underground. Significant water reserves will be required to synthesize propellant for Starship's return flight and to support normal colony operations (sanitation, agriculture, construction, manufacturing etc). Unfortunately this produces a dichotomy because water abundance tends to increase towards the poles, where the lower temperatures allow greater accumulation of ice. Large reserves of water could still be found at more median latitudes, although the likelihood of such reserves decrease because of the generally higher temperatures found at the equator.

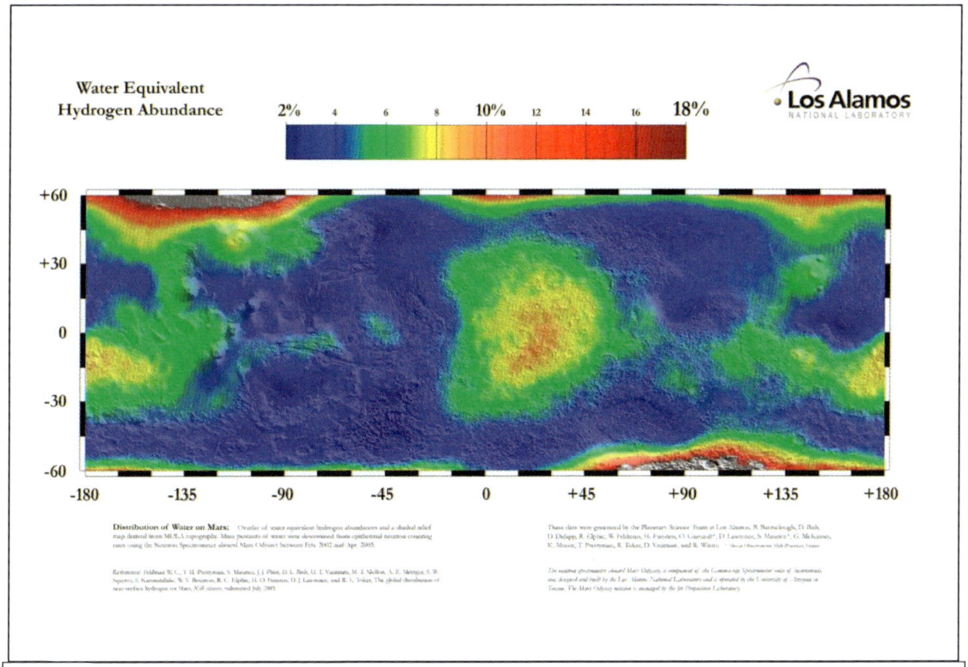

Water Abundance on Mars (at median latitudes)

Ideally any potential landing site should also be located at a lower elevation because this would provide more moderate living conditions for Mars colonists, less ambient radiation (deeper/denser atmosphere absorbs more solar and cosmic radiation) and should allow a higher tonnage of cargo to be delivered per flight (denser atmosphere provides more atmospheric braking, reducing terminal velocity, allowing a greater mass to be landed with the same amount of fuel). SpaceX are working with JPL to select the best sites that conform to all of these criteria i.e. high water abundance, median latitude and low elevation. Initial reports suggest the leading candidates for landing sites are: Deuteronilus Mensae, Phlegra Montes, Utopia Planitia and Arcadia Planitia(1).

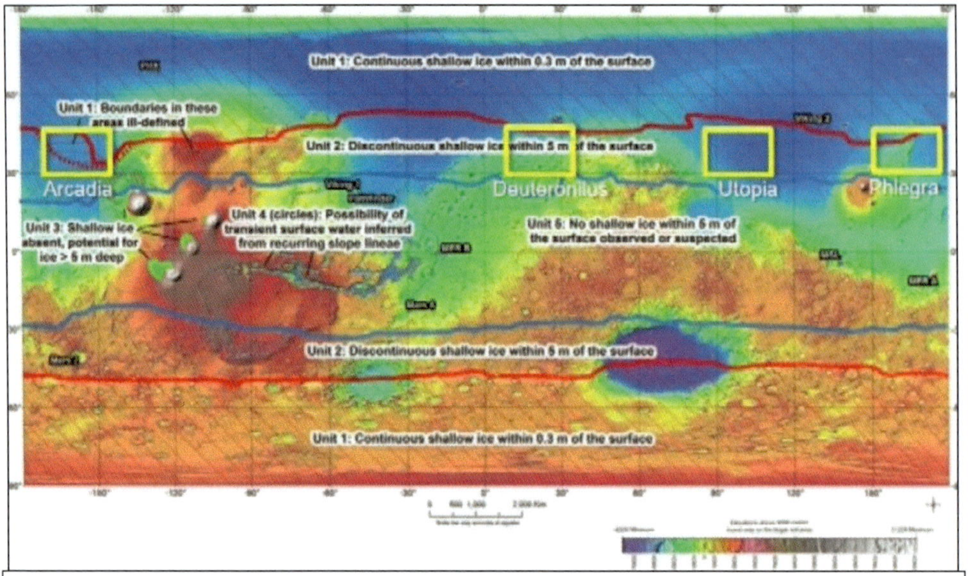

Topographic map of Mars with potential landing sites outlined

Arcadia Planitia (Mars coordinates: 47.2°N, 175.7°W)

Arcadia Planitia is located in the northern lowlands which are between 1 to 3 kilometers lower elevation than the adjacent southern highlands. The terrain consists of ancient volcanic plains which are relatively flat and uncluttered by boulders, potentially making it an ideal landing spot. Interestingly the area also possesses polygonal textured terrain, which resembles the landing area used by the successful Phoenix robotic spacecraft. These polygon blocks were likely formed when ice in the soil contracted as surface temperature fell (similar polygonal terrain features have been observed on Earth in polar permafrost and at high elevation). Fortunately Phoenix was able to confirm the presence of water in the immediate subsurface(2) which suggests water could also be relatively abundant at Arcadia due to the similar terrain.

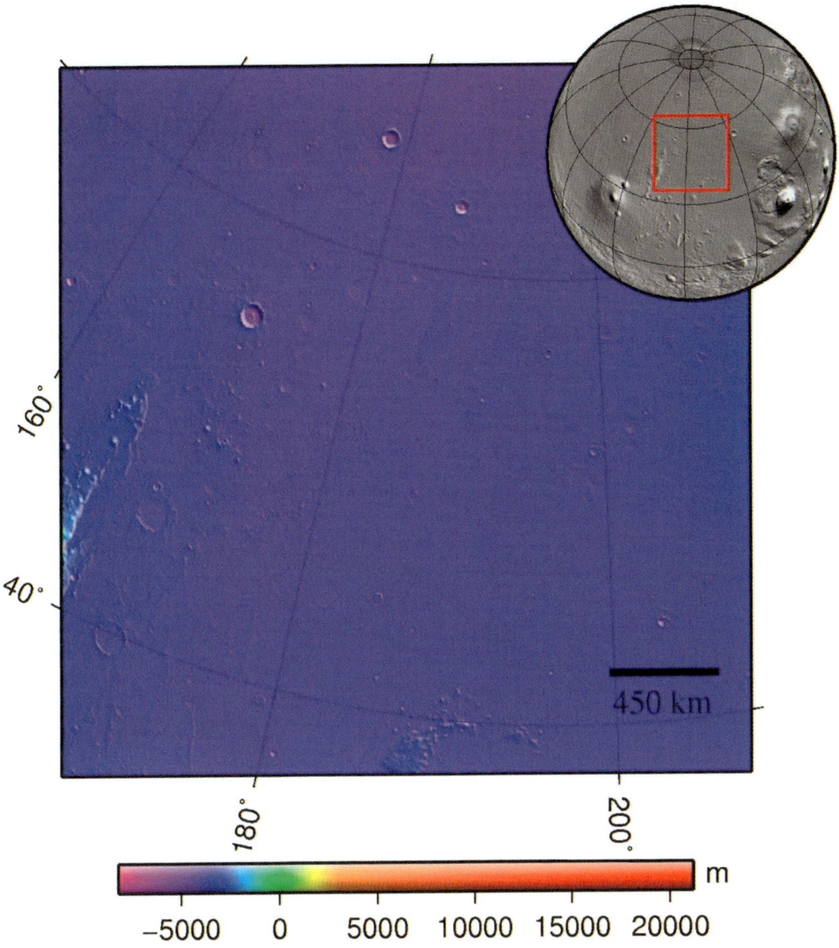

Topography map of Arcadia Planitia, credit Chmee2/Wikipedia

Arcadia Planitia is close enough to the equator to receive adequate sunlight throughout the year, which is essential for agriculture and solar power. Meteoric materials should also be present and easy to extract due to reduced erosion and tectonic activity on Mars. Overall the comparatively moderate climate found in these expansive lowlands could make it a great location for the first city of Mars.

Arcadia Planitia overall rating: 8 out of 10

Deuteronilus Mensae (Mars coordinates: 43.9° N, 22.6° E)

Mesa, glacier erosion and polygon terrain at Deuteronilus Mensae

Deuteronilus Mensae borders the southern highlands and the northern lowlands, hence possesses a comparatively low elevation mixed terrain. The Mars Reconnaissance Orbiter detected water ice in this area and there is evidence that glaciers persist here in modern times. Polygonal terrain has also been observed (which suggests an abundance of water in the subsurface) although these polygon structures appear distinctly domed which might prove hazardous for landing. Generally the area seems relatively rugged with ribbed features, brain terrain, mesas, deep surface cracks and pits. Cratering is fairly ubiquitous hence meteoric materials, such as rare earths and heavy metals, could be readily available and relatively easy to collect, via surface excavation.

Deuteronilus Mensae overall rating: 7 out of 10

Phlegra Montes (Mars coordinates: 40.4° N and 163.7° E)

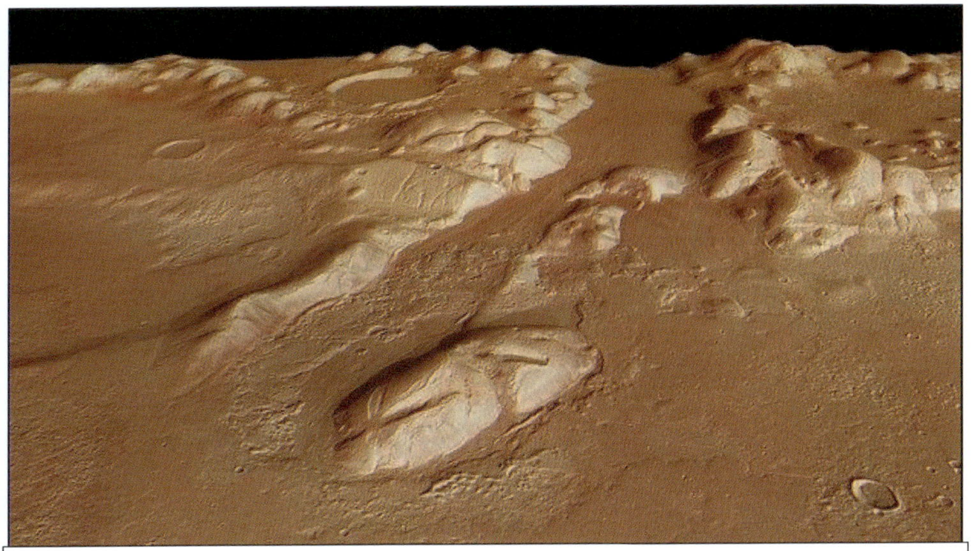

Glacial flows are prevalent at Phlegra Montes

Phlegra Montes is a 1,300 km long ridge of mountains which angle to the north-east across the northern plain. Strong evidence exists of glacial activity all around this ridge and the Mars Reconnaissance Orbiter has confirmed an abundance of water in the region. The areas of comparatively new terrain appear relatively smooth, with possible ice sheets located 20m under the rock debris surface. The Phlegra ridge was likely raised by tectonic activity which suggests some mantle minerals might also be found in this area. Phlegra Montes rises up to 4km above the surrounding northern plain, which presents a predicament. Agricultural domes located on the south facing slopes should be marginally more productive, due to the improved aspect to the sun. However, the high mountain ridge could constitute a hazard for initial landings, because of the lack of radio beacons or GPS tracking to guide their approach.

Phlegra Montes overall rating 7 out of 10

Utopia Planitia (Mars coordinates: 46.7°N 117.5°E)

Polygonal and scalloped terrain found at Utopia Planitia

Located in the mid Northern Plain the Utopia Planitia basin is the largest impact crater in the solar system, measuring 3,300 km in diameter. Its lobed, scalloped, cracked and polygonal terrain have long suggested the presence of subsurface water (lobes are evidence of glaciers and scallop patterns were likely formed when subsurface ice sublimated into the atmosphere leaving a ground depression). In 2016 the Mars Reconnaissance Orbiter confirmed large quantities of frozen water in the area, comparable in volume to Lake Superior, buried approximately 1 to 10 meters underground(3). In earlier times the basin likely contained a lake or small sea, hence sedimentary or water soluble materials should be relatively abundant, in addition to any meteoric materials derived from the impact basin's formation and subsequent bombardment.

Overall the region's exceptionally low elevation and open unobstructed plain certainly recommends it as a possible landing site.

Utopia Planitia overall rating: 8 out of 10

Landing Site Selection

A SpaceX spokesman reported that all candidate areas appeared viable when viewed using the medium resolution camera (CTX) on the Mars Reconnaissance Orbiter. However, when examined in more detail using the High Resolution Imaging Science Experiment (HiRISE) camera, three locations: Deuteronilus Mensae, Phlegra Montes and Utopia Planitia appeared to be much rockier.

"The team at JPL has been finding that, while the areas look very flat and smooth at CTX resolution, with HiRISE images, they're quite rocky. That's been unfortunate in terms of the opportunities for those sites." ~ *Paul Wooster (Spacecraft GNC Manager for SpaceX)*

Fortunately Wooster was able to confirm that Arcadia Planitia, while formed from lava sheets in more contemporary times, appears relatively smooth which makes it an almost ideal landing area.

"What they've found is basically few or no rocks and a polygonal terrain that they think is pretty similar to what was seen at Phoenix."

Arcadia Planitia landing sites

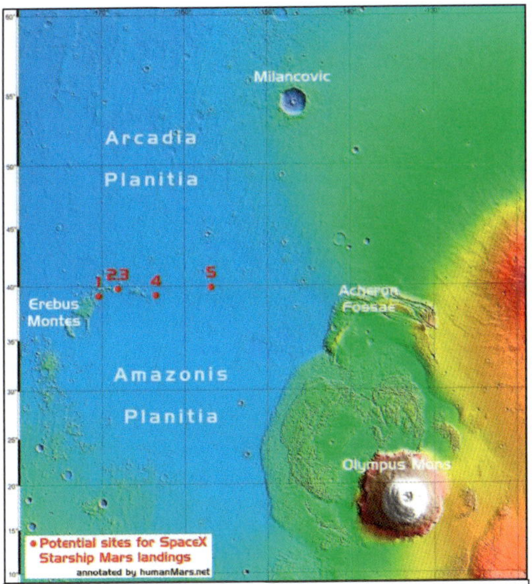

Five landing sites on the southern border of Arcadia Planitia

It has been reported that SpaceX are considering five possible landing sites, all located to the east of Erebus Montes in southern Arcadia Planitia(4). Interestingly, sites 2 and 3 appear to be virtually the same location, perhaps indicating increased interest in this area. There's evidence of underground water throughout this region, particularly on the downslopes of Erebus Montes. The landing area is in a volcanic floodplain, likely loaded with useful materials and Erebus Montes is within easy reach, which promises even more resources. Overall the climate's quite moderate (for Mars) and the potential landing area is vast, in case they miss their intended landing coordinates. Curiously Olympus Mons (the largest volcano in the solar system) is situated close-by, which could become a great tourist attraction!

CONCLUSION

It's possible that after further examination and analysis a suitably clear landing area might be discovered at an alternate site. However, SpaceX appear to be focusing on Arcadia Planitia as the primary landing location for their first scouting missions. Assuming these advance missions are successful, it seems likely they will progress to landing Mars colonists at (or near) the same location, during the next planetary conjunction window.

[1]http://spacenews.com/spacex-studying-landing-sites-for-mars-missions/
[2]https://en.wikipedia.org/wiki/Phoenix_(spacecraft)#Presence_of_shallow_subsurface_water_ice
[3]https://www.jpl.nasa.gov/news/news.php?release=2016-299
[4]https://behindtheblack.com/behind-the-black/essays-and-commentaries/spacex-begins-hunt-for-starship-landing-sites-on-mars/#more-60414

Chapter 12: First Mars Colony

MARS ARCHITECTURE OVERVIEW

As previously discussed, any Mars habitat would need to be a permanent structure and preferably substantial. Larger settlements are generally more viable because they support more people, who then provide a greater diversity of skills and present a broader gene pool. A large settlement is also less likely to be abandoned, due to vagaries of Earth politics, economics or commercial issues. So a large town would be the minimum requirement for a Mars settlement – preferably a small city.

"I think once we do establish such a city [on Mars] there will be a strong forcing function for the improvement of spaceflight technology that will then enable us to establish colonies elsewhere in the solar system and ultimately extend beyond our solar system(1)." ~ *Elon Musk*

In other words building a city on Mars should act as a catalyst for continued space exploration, allowing humanity to become a true spacefaring civilization.

It might seem strange to rely on private concerns, like SpaceX and associate companies, for such a momentous undertaking. However, most large construction, manufacturing and finance operations on Earth are routinely handled by commercial entities because they provide the most cost efficient and practical way to supply these services. Somewhat surprisingly, we shouldn't depend on NASA for such an undertaking because building a self-sustaining colony on Mars would likely exceed their space research remit and financial resources. NASA's charter directs them to perform long-range space studies, which is something they would no doubt choose to carry out at a Mars colony; so effectively they would engage with and support any Mars settlement without being its prime mover.

"The establishment of long-range studies of the potential benefits to be gained from, the opportunities for, and the problems involved in the utilization of aeronautical and space activities for peaceful and scientific purposes." ~ *NASA Charter/National Aeronautics and Space Act of 1958*

MARS TRAILBLAZER MISSIONS

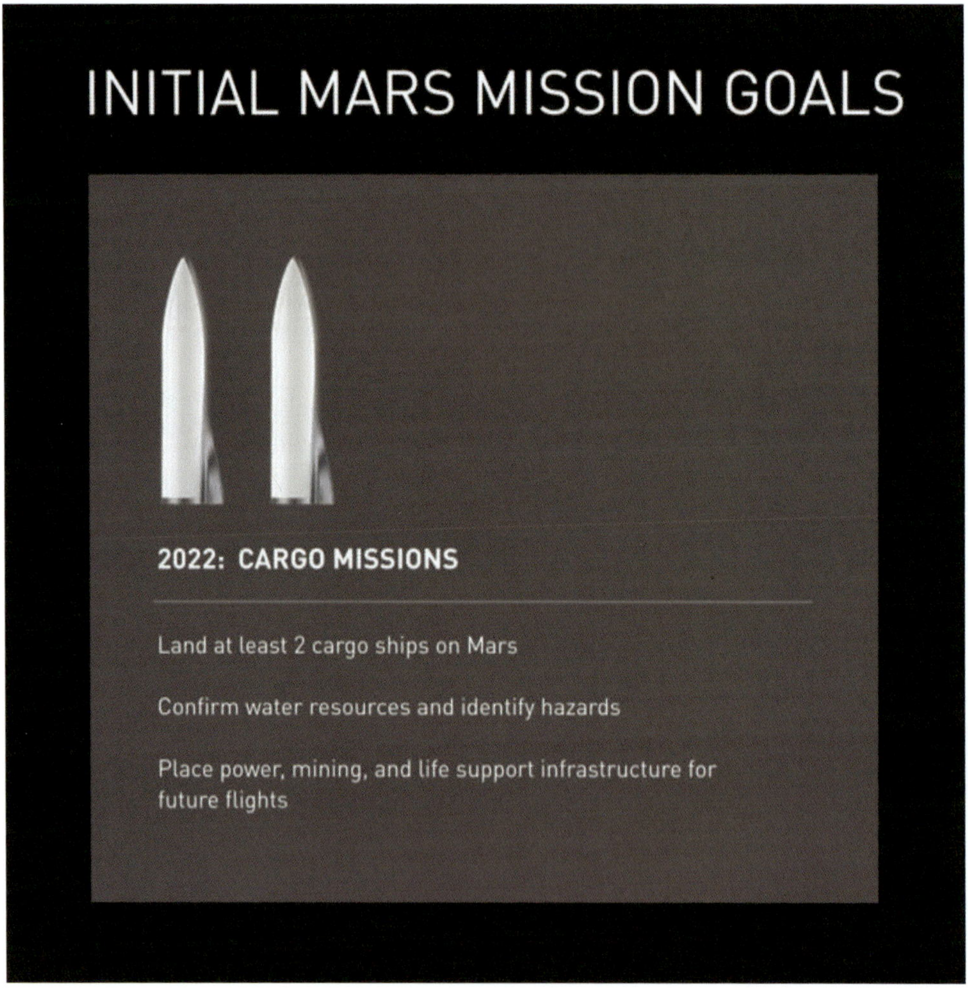

INITIAL MARS MISSION GOALS
2022: CARGO MISSIONS
Land at least 2 cargo ships on Mars
Confirm water resources and identify hazards
Place power, mining, and life support infrastructure for future flights

SpaceX have announced they intend to send at least two precursor missions to Mars in 2022(2). These unmanned Starship spacecraft should launch from Cape Canaveral and/or Boca Chica then take roughly 8-9 months to complete a one way trip. The primary function of these missions is to practice EDL procedures like aerobraking, supersonic retropropulsion and propulsive landing. If either landing is a success, they intend to use these craft to confirm there is sufficient water in the area for propellant production and identify any hazards for future missions. These spacecraft will also carry power generation, mining and life support equipment needed for the subsequent manned missions, which are due to launch during the next Earth-Mars conjunction window.

FIRST COLONY MISSIONS

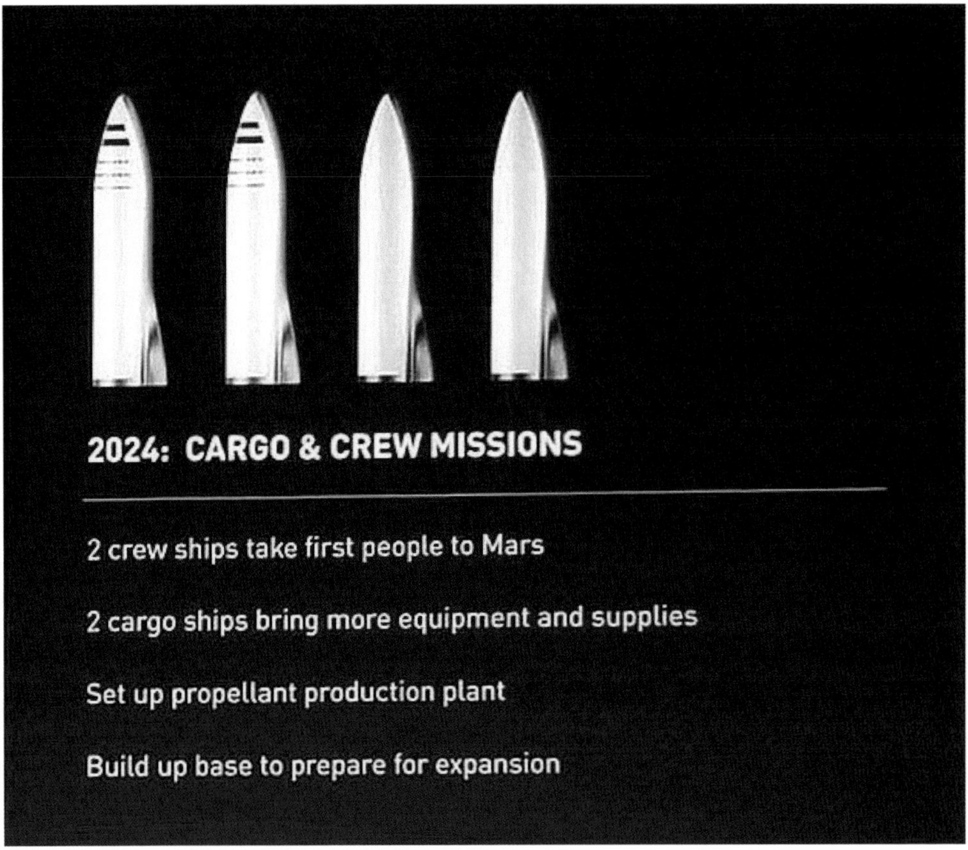
2024: CARGO & CREW MISSIONS
2 crew ships take first people to Mars
2 cargo ships bring more equipment and supplies
Set up propellant production plant
Build up base to prepare for expansion

SpaceX will need to import a lot of infrastructure to support a 2 year stay on Mars. This will probably require another 4 Starship flights (so six in total including 2022 missions), carrying multiple sets of equipment and a reasonable number of multi-skilled engineers. Going by past record, there is a good chance some of these vehicles will fail to arrive or land too far from the landing zone for viable use. Hence redundancy is key to everything; as the military say: "two is one – and one is none."

MARS COLONY: PHASE 1 (aka Mars Base Alpha)

Population: ~12(3)
Target Date: 2024(4)
Organization Structure: two tier meritocracy

"Early missions will be heavily weighted towards cargo. First crewed mission would have about a dozen people, as the goal will be to build out and troubleshoot the propellant plant and Mars Base Alpha power system(3)." ~ *Elon Musk*

FOUNDING A COLONY ON MARS

No doubt they will endeavor to make the Mars settlement as self-sufficient as possible due to the difficulties of supplying everything from Earth. However, much of the necessary materials, even structures, will need to be brought from Earth to build the first Mars colony. Here's a breakdown of the missions required to establish a Phase 1 settlement, suitable for 12 people.

First Starship Missions (circa 2022)

The first Mars colony missions will consist of two or more unmanned Starship vehicles, primarily used to prove the Mars Entry, Descent & Landing (EDL) techniques. If these automated landings are successful, they could deliver 200 mt of mixed cargo consisting of survey equipment and infrastructure materials for colonists on the following Starship mission. Here's a list of the cargo they'll probably carry to Mars on these first missions: -

1. Mars Rovers – SpaceX will probably send robotic rovers(5) to scout landing locations and ensure the ground is suitably clear/solid with sufficient reserves of subsurface water. These rovers could also be used to test whether ISRU propellant production is feasible on Mars (the following manned mission will require large quantities of ISRU propellant for Earth return). Multiple Mars rovers could be sent per Starship, stored in aft cargo bays. These surface vehicles should be capable of operating semi-autonomously as they survey the surrounding area.

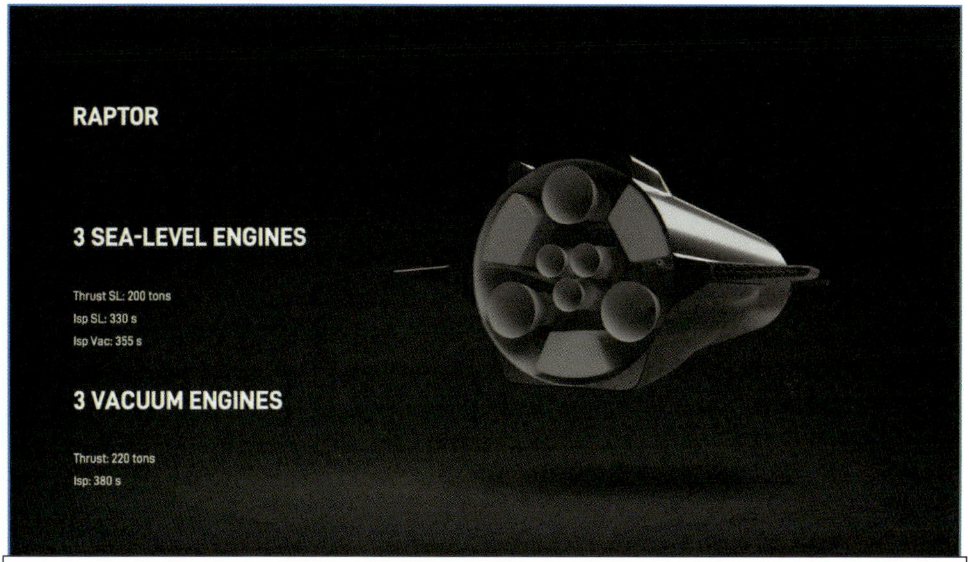

Starship's aft cargo bays are situated between the vacuum engines, credit SpaceX

"The logical thing to do is basically outfit one of the ships as a propellant plant itself, and just land it on the planet as a working propellant plant. And then you just need little miner droids to go dig up ice and bring it back and unfurl the solar panels(6)." ~ *Elon Musk/Popular Mechanics*

Elon Musk disclosed Tesla have developed a custom AI chip to allow their electric vehicles to perform Full Self Driving(7). This vehicle specialized chip (called the FSD processor) could also be used on Mars rovers, due to its low power consumption (estimated at less than 100W) and high processing capacity (144 trillion operations per second). This should allow them to field a next generation AI rover, capable of working quickly and with minimum human oversight (to reduce the effect of communication delay or dropout). If FSD is unsuited to the rigors of Mars off road driving, Tesla is reportedly developing an even more advanced FSD processor with three times the processing power, which should be ready in 2021 (approximately a year before the first Mars mission).

2. Surface Solar Arrays – Inflatable solar arrays seem a distinct possibility(5). These would provide power for the first mission and later on used as auxiliary power for the settlement.

3. Life Support Equipment – Before colonists arrive it will be vital to have spares in place to allow them to quickly repair any faulty life support equipment. Conceivably this might include a complete emergency habitat, in case the primary habitat (i.e. Starship) fails.

4. Mining Machinery – One or more mining machines(4)(8) will be required to extract water from the sub-soil. It seems likely these will be tunneling robots capable of fully autonomous operation. These advanced tunneling machines could also be used to excavate underground areas for industrial operations(8). Pressurized underground galleries will likely extend far beyond the colony's surface footprint and could be relatively large because of the inherently low gravity on Mars. Likely these underground workshops will constitute the first time in situ materials are used to construct colony infrastructure.

5. Mars Satellites – good satellite communications are vital for Mars operations and these satellites could also be used to improve landing accuracy, similar to GPS here on Earth. Hence it's possible some Mars modified Starlink satellites may also be deployed by these early cargo flights. This would require Starship to take a shallower approach to Mars and use the thin atmosphere to aerobrake to orbital velocity then lift up to low Mars orbit, (in a process called: "aerocapture"). Then a large number of these small satellites could be deployed from Starship's cargo hatch or rear cargo bays and left to maneuver to the desired orbits under their own propulsion. After deployment is complete, Starship would perform a deorbit burn and begin a second atmospheric entry before landing, something made possible by Starship's reusable design(9). Likely it will require multiple such flights to complete a functional Starlink constellation for Mars, ideally something they would like to have in place prior to crew landings.

Manned Starship Mission (circa 2024)

Assuming the automated test missions are successful, the first manned Starship flights will aim to launch two years later during the next planetary synchronization window. These craft will carry around twelve(3) of SpaceX's finest with enough provisions and ground equipment to support a two year mission. Here's a list of the equipment they intend to take to Mars to support this first settlement: -

1. Propellant Plant – One or more propellant plants will be needed to refuel the return Starships. Once set up, significant quantities of methane and oxygen could be synthesized within each plant, using mined water and CO2 derived from the Mars atmosphere.

2. Human Habitation – A heated/pressurized sanctuary from Mars surface conditions which could comfortably accommodate at least twelve people. This surface habitat is likely to be a geodesic dome design, consisting of a carbon fiber frame with glass panels. Likely this will serve as dual function habitat and agriculture module, providing a green self-oxygenating living space(8).

3. Fertilizer Plant – Fortunately SpaceX can produce fertilizer from the Martian atmosphere, which is 2.7% nitrogen(10). A relatively small machine could produce all the colony's fertilizer requirements, using in situ resources.

5. Surface Solar Arrays – Large solar arrays will supply the power for mining, atmospheric processing and propellant production(4). Solar panels will likely be transported as flat pack cargo, allowing them to be easily offloaded and unpacked by colonists. It's likely these high efficiency arrays would charge high density batteries, which could potentially provide most of the settlement's energy needs.

"So we should, particularly with 6 ships, have plenty of landed mass to construct the propellant depot, which will consist of a large array of solar panels – very large array, and then everything that['s needed] to mine and refine water, and then draw the CO2 out of the atmosphere, and then create and store deep cryo CH4 and O2(4)." ~ *Elon Musk/IAC 2017*

6. Nuclear Reactor – A nuclear power source will be required(11) as a backup generator (dust storms can last a week or more on Mars and completely clog any solar arrays). It's rumored this reactor will be 20 metric tons, 6 m tall and 5 m diameter(12). Likely it will operate at night when there is no power available from solar arrays. Auxiliary nuclear power is an unfortunate necessity on Mars, the settlement would certainly struggle to survive in the event of a complete power failure.

"You need a nuclear reactor for power, heat, the ability to melt water, the ability to do electrolysis on the water to get oxygen out(11)." ~ *Adam Lichtl, former Director of Research at SpaceX*

The Starship spacecraft need to be refueled before they can depart Mars, hence SpaceX colonists will have to setup a propellant manufacturing plant delivered on the Starship cargo vessels(3)(5). This plant should combine carbon dioxide from the atmosphere with water from the soil to create thousands of metric tons of methane and oxygen propellant. The process should be highly automated, with relatively low power consumption, allowing the necessary propellant to be manufactured during the two year interval between Earth launch windows. Power for this ISRU propellant plant is primarily supplied by solar collection.

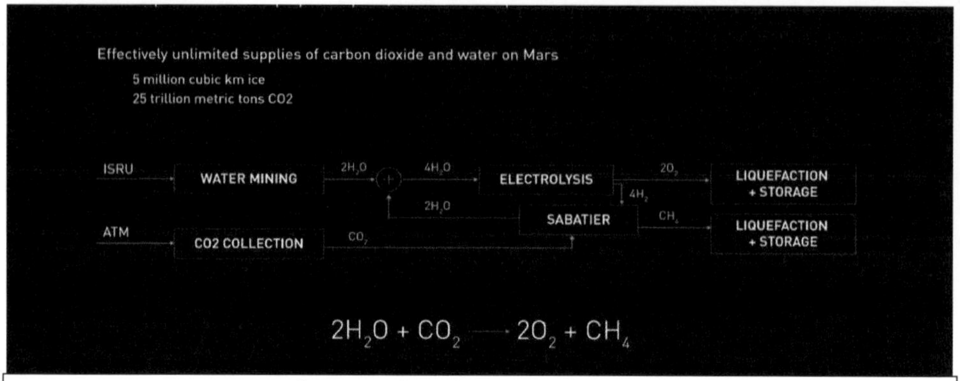

Propellant manufacturing process, credit SpaceX

It is possible this ISRU propellant will be stored directly in the Starship spacecraft's inner propellant tanks, which should be capable of long term propellant storage. This would allow colonists to use a single Starship as a rescue vessel(13), after sufficient propellant is combined into one spacecraft.

Phase 1 Culture

Early Mars settlers will express a markedly different culture to Earth. These engineers will be committed to living on Mars and fully aware of the sacrifices they need to make to survive. Their sole purpose for being on Mars is to do whatever's necessary to establish the colony's infrastructure. In less than two Earth years they will have to ensure all equipment is unpacked, installed and operational (from all Starship missions) – before the first wave of explorer/scientist/entrepreneurs arrive, during the next planetary conjunction.

Likely this will lead to a relatively simple hierarchy of values: -

1. Preserve human life and well-being
2. Preserve infrastructure of the settlement
3. Conserve food production/reserves
4. Conserve material resources

Preserving the well-being of these early colonists will be paramount, systems might fail and need to be replaced but there's no way to replace people. These multi-skilled engineers would be of extraordinary high value in this minimalist survival setting. The close interdependence of individuals will no doubt foster strong teamwork. Likewise, a strong work ethic will be universal, in fact competition over "who works hardest" could be the norm. SpaceX intend to send many Starship vessels, which should effectively supply a surplus of infrastructure and food production equipment for the settlement. Hence these resources should be vital but not as important as people. Material resources (water, air, computers, survival suits etc) should also be oversubscribed, hence of lesser importance than life, infrastructure and food.

Mars Economy

At this stage, money will have no meaning on Mars, the only currency will be survival. Salaries accrued on Mars could be spent on Earth to buy virtual services: e.g. software, entertainment, financial products etc but anything these pioneer engineers could possibly need would be provided them, given the constraints on transport. Any exchange of goods or services between individual colonists would probably use barter for payment; essentially money would have little to no practical value in this survivalist setting.

Colony Autonomy

From the outset colonists will have a high degree of autonomy, some might even argue independence. They have put their life on the line to spearhead humanity's evolution into a spacefaring civilization and perform herculean tasks daily to help establish the first offworld settlement. Any attempt to influence or persuade them to do anything that distracts from this task, puts them at further risk or simply makes their life any harder would be poorly received and quite possibly ignored. Effectively the desires of Earth would come second to the colonist's more imminent and pressing needs. Probably the best way to illustrate this effect is through exploring some practical examples of colonist responses to distant Earth directives: -

Earth directive	Colonist Response
Collect regolith samples from numerous locations many days travel from the colony, ignoring any additional radiation exposure	Ensure samples from boring machines are continually monitored to minimize any chance of contaminants entering the colonies ISRU plant and affecting its water supply
Construct surface domes as a priority to provide a perfect photo opportunity for public relations	Expedite the construction of underground tunnels which can double as radiation shelters in case of solar storms
Refrain from personal relations and observe sexual abstinence to ensure the best professional attitudes are maintained, with minimum distraction	Pursue committed relationships for mutual support and emotional security to produce a stronger and more durable society
For surface operations, personnel, suits and airlock must be scrubbed both before and after each excursion to minimize any chance of contamination	Conserve the water to share a water saving shower with loved ones

Generally, any highhandedness by ground control won't wash... Once colonists land they'll want to make their own decisions, they're certainly in the best position to judge; It's their boots on the ground and head on the block.

Phase 1 autonomy rating: 8 out of 10 ("Thanks for the advice...")

CONCLUSIONS

1. Achieving their first Mars landings in 2022 is an ambitious goal for SpaceX but still feasible considering they have been planning and preparing this for 15 years. A case of light the blue touchpaper and stand back.

2. Establishing a Phase 1 settlement is possible with current technological understanding, however, such an undertaking could severely stretch SpaceX resources. Hence it's entirely possible these efforts will be fostered by NASA and other international partners, once SpaceX have proven their Starship LS design. Currently NASA has its own plan to visit Mars, using the Space Launch System (SLS), so might avoid engaging with SpaceX until the manifold advantages of Starship become manifest (see Appendix 3: Mars Mission Comparison). For example, an SLS Mars mission would have a relatively small crew, less equipment and no doubt considerably more expensive, because it relies on expendable technology. In addition a SLS Mars mission would produce only a fraction of the ubiquitous advances derived from long term settlement.

To be succinct: We know how to boldly go – and must now learn how to boldly stay.

[1]https://youtu.be/pIRqB5iqWA8?t=1723
[2]https://youtu.be/S5V7R_se1Xc?t=2226
[3]https://www.reddit.com/r/spacex/comments/590wi9/i_am_elon_musk_ask_me_anything_about_becoming_a/d94txm0/?context=3
[4]https://www.reddit.com/r/spacex/comments/73cw1u/my_notestranscript_elons_iac_2017_talk_parts_1_2/
[5]https://youtu.be/4DUbiCQpw_4?t=1653
[6]https://www.popularmechanics.com/space/moon-mars/a26513651/elon-musk-interview-spacex-mars/?utm_source=reddit.com
[7]https://electrek.co/2019/04/23/nvidia-disputes-tesla-fsd-computer/
[8]https://www.reddit.com/r/spacex/comments/590wi9/i_am_elon_musk_ask_me_anything_about_becoming_a/d94t2bv/?context=3
[9]https://youtu.be/zu7WJD8vpAQ?t=6024
[10https://youtu.be/wB3R5Xk2gTY?t=3619
[11]http://www.smmirror.com/articles/News/Inside-SpaceX-Getting-From-The-Blue-Marble-To-The-Red-Marble/44073
[12]http://www.spaceflightinsider.com/organizations/space-exploration-technologies/spacexs-mars-colonial-transporter-rumors-realities/
[13]https://www.reddit.com/r/spacex/comments/590wi9/i_am_elon_musk_ask_me_anything_about_becoming_a/d94voyh/?context=3

Chapter 13: Lean Green Mars Making Machine

When the initial Mars colony is established, with all necessary ground equipment operational, SpaceX can move to phase 2 settlement. Ideally the process of establishing a more advanced colony on Mars could be left to some third party, leaving SpaceX free to focus on space transport. However, finding anyone willing to invest in such a divergent undertaking might prove interesting. Likely SpaceX will need to establish a relatively robust colony on Mars before they can attract major investors. Essentially they will need to supply somewhere worth travelling to before major commercial investors become significantly involved. This will likely require SpaceX to build at least a Phase 2 settlement and responsible for a fair proportion of Phase 3, the first city of Mars.

MARS COLONY (PHASE 2)

Population: 100+(1)
Date: 2028(2)
Organization Structure: three tier technocracy(3)

A small fleet of Starship spacecraft will transport large bio-domes for human habitation(4). These could be similar to the geodesic dome used in Phase 1 except on a much grander scale. Likely these structures will become increasingly specialized, with separate accommodation, agriculture and civic centers.

The number of Starship spacecraft shuttling between Earth and Mars should increase during each planetary conjunction window. Also, the proportion of passengers to cargo transported will increase when all primary settlement infrastructure is in place (pressure domes, surface vehicles, power generators, propellant plants, landing zones, additive manufacturing shops, organic and inorganic recycling facilities etc). Nearly everything will need to be recycled, including air and human waste. Replacing any resource would be a magnitude more difficult on Mars, particularly if sourced from Earth.

Phase 2 Culture

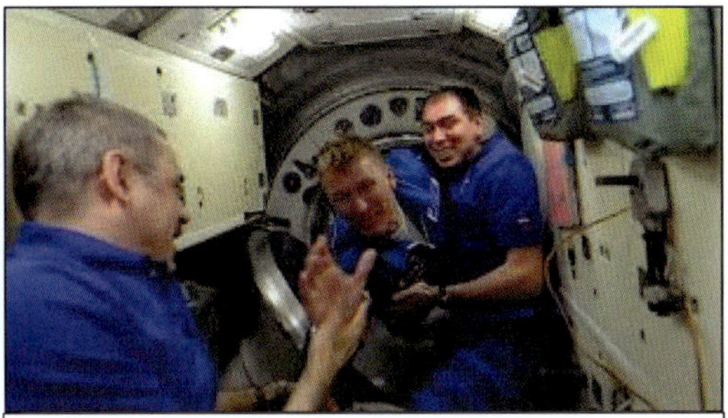

Astronaut Tim Peake is welcomed aboard the International Space Station

Likely the Phase 1 engineers will be widely regarded as ideals by fellow colonists, following their heroic work to establish the settlement. This group will have demonstrated the importance of teamwork, mutual respect and collaborative survival under the most challenging conditions, incidentally, helping to form a new social outlook. This new perspective can perhaps be glimpsed by the rapturous way they greet new arrivals on the International Space Station. Anyone who arrives at the ISS has put everything on the line for humanity – and the person standing right next to them; they have literally passed through fire to get there. No doubt this new social philosophy will undergo many trials with the influx of new arrivals on Mars.

In phase 2 a steady stream of engineers, explorers and entrepreneurs will arrive to supplement the colony's population. To progress from Phase 2 to Phase 3 the populace will need to be healthy, long lived and happy in order to achieve the necessary high productivity. Hence people will remain prized above anything, even infrastructure, which should become more abundant as the Starship fleet steadily grows. Similarly, food production/reserves should also be relatively high, which will likely mean the material resources required to expand the settlement take priority. Each Starship transport will supply an increasingly large proportion of people compared to cargo and every one of these new arrivals will need resources. Hence the colonies hierarchy of values will likely transform to become: -

"[Mars needs] Engineers, artists & creators of all kinds. There is so much to build." ~ *Elon Musk*

1. Preserve human life and well-being
2. Preserve the infrastructure of the settlement
3. Conserve material resources
4. Conserve food production/reserves

Mars Entrepreneurs

"...that's [Mars] really where a tremendous amount of entrepreneurship and talent would flourish, just as happened in California when the Union Pacific railroad was completed. And when they were building the Union Pacific, people said it was a super dumb idea because 'there is hardly anybody living in California.' But now today we have sort of at least, U.S. epicenter of tech development and entertainment and it's the biggest state in the nation(5)." ~ *Elon Musk*

Likely Mars will become famous for its entrepreneurs, who will certainly go a lot further than their earthbound equivalents. These will likely range from the smallest service providers to full scale resource extraction. For example, levels of atmospheric methane have been observed to rise two or three times during the spring and summer seasons. These sporadic rises suggest there might be underground traps of methane which vent to the atmosphere as sub-surface ice melts in warmer conditions. If entrepreneurs can locate these potentially large reserves close to the surface, they would likely have a ready-made market for Starship propellant. Extracting methane from the ground should prove a magnitude less expensive than the alternative i.e. synthesizing from CO2 and water which is highly energy intensive.

"We observed and mapped multiple plumes of methane on Mars, one of which released about 19,000 metric tons of methane(6)" ~ NASA

It seems, even at this early stage entrepreneurs are lining up to begin operations on Mars. Relativity space plan to produce the first rockets there using their unique 3D printing technology. No doubt operating a quickly configured and constructed launch vehicle might be of great value to other space orientated entrepreneurs, as the colony continues to grow.

"Relativity was founded with the long term vision of 3D printing the first rocket made on Mars and expanding the possibilities for human experience in our lifetime(7)." ~ *Tim Ellis, Relativity CEO*

Mars Economy

A Mars (electronic) currency will likely appear around this time although of only marginal use. Healthcare, food, air, water and all reasonable needs will be provided for, free and gratis (SpaceX employees already have free healthcare, administered in the workplace). Mostly this virtual currency will be used to purchase commodities (like surface rovers, portable habitats, increased bandwidth, specialist apps etc) and import luxury goods from Earth. During Phase 2 private ventures will increasingly arrive on Mars, many attracted by the substantial sums of government money spent by Mars explorers and scientists. The education sector should also begin to play an important role in the Mars economy. Increasing numbers of anthropologists, architects, historians etc will arrive to study the new Mars colony. Many of these engineers and academics could decide to stay and help build the first Mars educational institutions, which should strongly influence Phase 3 development.

To ensure no one slips between the cracks, all citizens will likely receive 'basic' i.e. a minimum basic income from Mars authorities. If colonists lose their work due to automation, suffer long-term debilitating injuries or need to retire, Mars will be morally obliged to support them and provide a financial safety net. Hence basic would be awarded to everyone, no matter their circumstances. Essentially each colonist could choose to remain on basic (and pursue personal goals and interests) or work (i.e. contribute to the colony in some substantial way) in order to increase their income and supplement their basic allowance.

"There is a pretty good chance we end up with a universal basic income, or something like that, due to automation... People will have time to do other things, more complex things, more interesting things, certainly more leisure time(8)." ~ *Elon Musk*

Colony Autonomy

As the colony's income increases it will no doubt acquire an even more independent attitude. Any newcomers from Earth will have to play by 'Mars rules' when it comes to sharing the colony's precious resources (food, water, power etc). No doubt colonists will generally remain on friendly terms with new arrivals due to their continued reliance on Earth's financial support (similar to the attitude of locals at tourist hotspots). However, Earth will find it increasingly difficult to exert any real control over Mars citizens, who increasingly realize they have something Earth wants – which puts them in a strong bargaining position.

Phase 2 autonomy rating: 9 out of 10 ("Earth attitudes are quaint")

PHASE 2 CONCLUSION

Transferring goods and materials from Mars to Earth will be expensive on resources and quite possibly reduce the service life of Starship spacecraft. Hence exports from Mars will likely be restricted and quite possibly limited to samples of geologic materials. For this reason, the Mars economy will primarily be focused on supporting their own people and infrastructure, at least until there is some fundamental improvement in launch technology. However, the nascent Mars economy should be surprisingly energetic, Mars citizens and companies will no doubt require a comparatively large amount of homegrown goods and services, in order to maintain a reasonable lifestyle and expand the colony under some fairly challenging circumstances.

"[Mars Base Alpha should be self-sufficient after] About 10 orbit synch windows, which are every 26 months, so self-sufficiency around 2050(9)." ~ *Elon Musk*

MARS COLONY (PHASE 3)

Population: 80,000+(10)
Date: ~2100(10)
Organization Structure: technocrats empowered by direct democracy(11)

Larger habitation domes can be expected as settlement progresses, distributed over multiple areas. Separate population centers will likely be established for mining and terraforming programs. Generally the colony will become self sufficient with only luxuries supplied by Earth. Nuclear fusion or similar gigawatt scale power generation should now be available, allowing terraforming efforts to proceed at full pace.

'Kick-start Mars' Terraforming

Terraforming an entire planet is an enormous undertaking hence terraforming engineers will likely take the most direct and inexpensive route possible to get the ball rolling. To start the terraforming process, multiple technologies will likely be employed in parallel, in order to precipitate a runaway greenhouse effect. This technique might sound dubious, due to hazards we currently face on Earth from climate change. However, the average temperature on Mars is 70°C lower than Earth's, so the prospect of any increase in ambient temperature would probably be warmly welcomed by Mars colonists.

To start, the terraforming process will probably include large scale release of greenhouse gases in addition to melting of the polar icecaps using nuclear energy(4), which should add even more greenhouse gases (primarily carbon dioxide and water vapor) to the atmosphere. This ensures more of the sun's heat will be retained at the surface, which would further increase the amount of greenhouse gases evaporating into the atmosphere globally. Eventually this should trigger a runaway greenhouse effect, which would significantly raise ambient temperatures (by 20°C or more). Rain could fall for the first time in a billion years, forming streams and rivers that cascade down into great lakes dotted around the globe. Extremophile plants could then be used to liberate oxygen and retain even more heat from the sun(12) making the climate even more clement. Note colonists would still require sealed environment suits until a stable biome could be established and ironically this might become increasingly difficult to achieve as temperatures continue to rise.

Planetary scale production of greenhouse gases on Mars

Colonists could possibly encounter three increasing problems during the 'kick-start Mars' process. The radiation levels at the surface of Mars should reduce because the growing blanket of atmosphere absorbs more of the ambient solar and cosmic radiation. However, surface radiation should still be higher than on Earth because Mars lacks a robust magnetosphere (Earth's magnetosphere helps to deflect solar and cosmic radiation, effectively producing a radiation free oasis here on Earth). In addition this ionizing radiation, combined with increased heat from global warming, could produce considerable atmospheric disturbances, precipitating thunderstorms, tornados or possibly more extreme weather phenomena. A suitable example of this effect would be the turbulent atmosphere found on Venus, which also lacks a strong magnetosphere to protect it from ionizing radiation. Lastly solar storms, i.e. radiation storms originating from the sun, could take large scoops out of Mars's young hot atmosphere, repeatedly setting back the terraforming process by months or even years.

Problems like these should highlight the fact that 'kick-start Mars' has actually produced a zombie planet i.e. one which seems alive (like Earth) but is in fact in continual decay - and prone to bouts of tempestuous rage against the living!

Terra-pause

Colonists will probably want a hiatus from terraforming while they re-evaluate the available techniques and regain confidence in the terraforming process. During this interim period new technologies should be developed to embark on an even more ambitious terraforming project, designed to increase habitability and produce a 'living Mars'. The specific technology(s) required to control Mars's errant environment are difficult to imagine but let's look at one possible scenario, based on what we know.

Present day discoveries about Earth(13), Enceladus (a moon of Saturn) and the dwarf planet Pluto have proved there's a large heat source at the center of many worlds which drives their volcanic activity. The visionary scientist, J. Marvin Herndon, suggests this heat is generated by a naturally occurring fission reactor located inside the inner core of such bodies, called a georeactor(14). Around 4.6 billion years ago the earth coalesced from interstellar dust and gases. The heat generated by Earth's formation, decaying isotopes and asteroid impacts caused it to melt then differentiate into layers. Lighter elements like silicates rose to the surface to form a crust, while heavier elements, like iron, sank to form the core. However, the heaviest elements of them all, like uranium, sank even further to occupy the inner core. When uranium becomes sufficiently concentrated it usually initiates a nuclear reaction, similar to what occurs in manmade nuclear reactors. The heat from this reaction causes the liquid iron in the core to circulate, in a convection process, cycling between the hot inner core and cooler planetary crust. This circulation of conductive iron generates a protective magnetic field around the world called a magnetosphere. This magnetic field, which is many times larger than Earth, deflects the charged radiation from the sun (solar radiation) and background radiation from the Universe (cosmic radiation).

Interestingly Mars once possessed much higher core activity, which produced a strong magnetosphere and volcanic activity. The high volume of volcanic outgassing and strong magnetosphere helped to create and maintain a thick atmosphere and more importantly produced low surface radiation. At present the core activity on Mars has reduced to such a level that it can no longer circulate sufficient liquid iron to generate a strong magnetosphere. However, there is still sufficient heat produced in the georeactor to keep the iron fluid(15), just not circulating, so the georeactor is in what could best be described as 'standby mode'.

In theory, to produce a habitable Mars with long term stability, the georeactor core activity will need to significantly increase, to reproduce the natural process originally used to terraform Earth. Before such a staggering feat of planetary engineering could be attempted it seems likely that tests would be performed on some little habited spot, like the nearby protoplanet Vesta or dwarf planet Ceres (named after the Roman goddess of agriculture). If a barren rock like Ceres could be made to bloom, it should certainly help restore public confidence in the Mars

terraforming process. Uranium (likely sourced from the asteroid belt) would gradually be introduced into Ceres' core, allowing researchers to incrementally increase georeactor activity. The increased heat should produce large geyser-like eruptions and eventually allow researchers to manipulate sections of the crust, in a process similar to plate tectonic movement. Assuming some kind of stable magnetosphere can be established, the researchers should then be given the go-ahead to adapt these georeactor control techniques for use on Mars.

'Living Mars' Terraforming

So to complete the terraforming process, Mars colonists could attempt to increase the georeactor core activity to create a viable magnetosphere. Fortunately the large quantities of uranium required can mostly be mined on the surface of Mars. Currently there is negligible tectonic plate movement on the planet hence the remnants of all the asteroid impacts from the last billion years (or more) have steadily accumulated across the surface. Some of this asteroid debris should include high concentrations of uranium which could be relatively easy to mine because of the lack of water erosion.

Lava tunnels on Mars are larger than their equivalents on Earth

When the georeactor core activity originally decreased on Mars it seems likely the lava retreated back to the planet's fluid core. These dry lava tunnels could be used to transport large quantities of uranium 235 and nuclear fertile materials (like thorium and uranium 238) to a sufficient depth to deposit them directly into the molten core. These ultra-heavy materials should then sink to the planet's georeactor core and react with the pre-existing nuclear fissile materials. This should immediately increase core activity and markedly improve its heat output. Assuming the georeactor is carefully monitored and controlled this increased core activity should be stable and sustainable. A planetary georeactor is in effect a fast breeder reactor hence if sufficient fertile materials (like thorium) are available in the core, it can use them to breed enough nuclear fuel to operate almost indefinitely, similar to the stable process we see in Earth's georeactor core.

If sufficient uranium 235 and fertile materials can be added to the core, Mars's georeactor should eventually be restored to the high power output it enjoyed in the first billion years of its life, after the planet's formation. Increased heat output from the core should restart volcanic activity and tectonic plate movement. Subduction of these tectonic plates would act like a conveyor belt, transporting more nuclear materials from the surface to the fluid core, keeping the georeactor fed with fuel. The iron convection in the fluid core should generate a strong magnetosphere which extends thousands of miles into space. Then the entire atmosphere would be safely contained within a magnetic bottle which should also help to deflect solar and cosmic radiation.

In addition, pyroclastic outgassing should introduce more greenhouse gases into the atmosphere, further elevating global temperatures. Vast reserves of water, long trapped underground, would be forced to the surface as steam, hot springs and geysers, probably resulting in the formation of an ocean, stretched across the entire northern hemisphere.

Then rain will wash the pyroclastic gases out of the atmosphere to fertilize the soil. In other words life giving elements are continually cycled from the surface to the core and back again to the surface - the circle of life complete. As the final touch genetically adapted flora and fauna could be introduced to complete the terraforming process. Likely at the end of this unprecedented project they will have created a living breathing planet, much like Earth.

Living Mars Caution

It should be noted that the Living Mars terraforming process contains many diverse hazards and should be handled with the caution it deserves. If a mother lode of uranium was inadvertently ingested by a subduction fault, it could easily lead to excess volcanic activity and increased tectonic motion. Georeactors appear relatively stable and self-regulating, however, there is probably a limit to how much fissile materials can be added at once – quite possibly related to the planet's gravity.

Mars's twin moons, Phobos and Deimos, have similar composition to Mars, so it's possible some supercritical event in Mars's primordial georeactor could have thrown part of the planet's crust into orbit to create these twin companions. A supercritical nuclear reaction occurs when fissile materials become too concentrated, causing a runaway chain reaction, resulting in a nuclear detonation. There is some evidence this might have occurred on Mars during its primordial period, around a billion years after it was formed. Fissile materials were a lot more abundant then because highly reactive uranium tends to decay over time into less reactive 'daughter' elements.

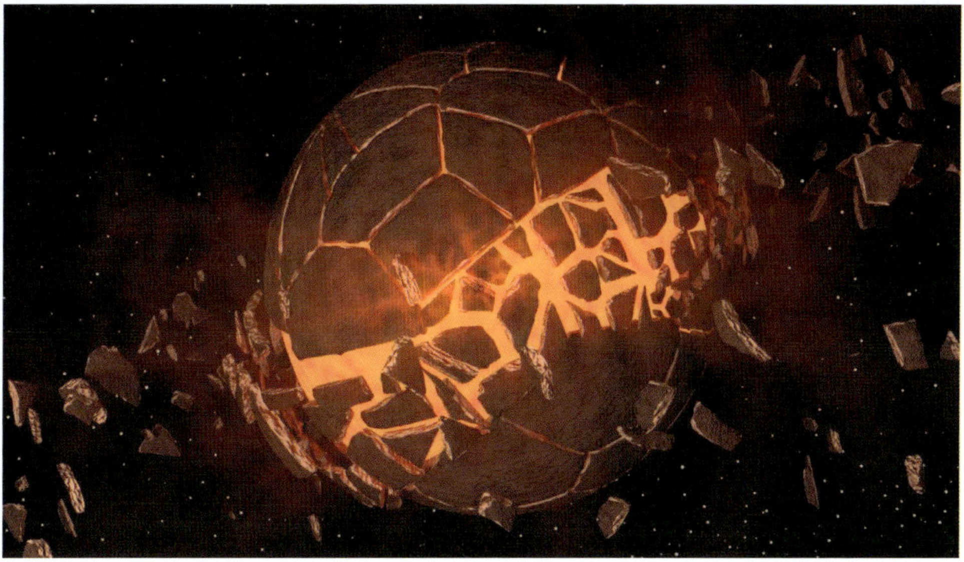

We know there was some kind of cataclysmic event early in Mars's life which entirely altered its geography. The northern hemisphere of the planet has significantly lower elevation than the southern hemisphere, for example the southern highlands are 26 km higher than the northern lowlands(16) which is an inordinate difference in elevation. There are also deep fissures in the planetary crust found to the south, such as the Valles Marineris, which run up to seven kilometers deep. Perhaps the best evidence of some kind of nuclear event is the widespread surface deposits of uranium, thorium and radioactive potassium(17), spread across large areas of the northern lowlands (elements which are normally found in the georeactor core). In addition iron oxide powder is almost ubiquitous across the surface of Mars (normally such large deposits of iron would only be found in the fluid core) which gives the planet its characteristic red color. Last but not least, the planet experienced unprecedented volcanic activity during this cataclysmic period, which suggests the georeaction driving these eruptions was far from stable. Given the evidence it seems possible some kind of dysfunction occurred in the primordial georeactor which dispersed core materials across the surface and possibly hurled a large section of the northern planetary crust into space. Preventing such an extreme event in Mars's georeactor could prove the greatest challenge to these early terraforming engineers. Big projects mean big problems but this is the kind of challenge that world class engineers thrive on.

In addition to planetary mechanics problems, there might also be difficulties adapting flora and fauna to survive in the emerging environment on Mars. Genetic engineering technology could be unbelievably potent in this era hence caution will need to be taken not to make the genetically modified flora and fauna too adaptable.

Day of the Triffids...

Phase 3 Culture

It will be tough to raise a family on Mars and probably involve routine remedial treatment e.g. gene therapy to overcome problems associated with low gravity, low air pressure and higher than normal radiation. However, gene editing techniques should be commonplace in this era - even used for cosmetic purposes (e.g. designer genes).

"This (Crispr) or analogous approaches may one day enable human genome targeting or editing during very early development(18)" ~ *Tony Perry/The Guardian*

Use of in situ resources should now be ubiquitous hence people become the only thing of any real value (anything else could be readily replaced or fabricated by automata). Likely the average age of this settlement will steadily reduce as increasing numbers of juveniles arrive from Mars's twin moons and assorted mining concerns in the asteroid belt. These children will require adequate gravity and custom gene therapy to ensure strong development hence enrolled as school boarders on Mars for the duration. Earth gravity could prove fatal to children born in microgravity and Mars will likely be less expensive and more experienced in delivering remedial treatment. Many of these juvenile boarders will choose to stay, resulting in the colony achieving the mythical one million figure before the millennium's close.

Colony Independence

In Phase 3 Mars should become self sufficient/sustaining with a steadily expanding population – displaying marked different attitudes to Earth. If independence hasn't already occurred in late Phase 2, it should definitely arrive in Phase 3. Earth will likely acquiesce to this move to independence, considering Mars's emerging status as the hub of the space economy.

Phase 3 autonomy rating: 10 out of 10 ("Earth needs help")

PHASE 3 CONCLUSIONS

1. Terraforming Mars will be incredibly challenging and sophisticated and likely require a lot more effort than 'merely' melting the poles. However, once a viable terraforming process has been created, it could then be used to rapidly terraform other worlds.

2. If economic activity in space exceeds Earth, Mars will likely become the hub of human development.

[1]https://youtu.be/wB3R5Xk2gTY?t=2930
[2]https://twitter.com/elonmusk/status/1043253619485622272
[3]https://twitter.com/elonmusk/status/1142890265369202688
[4]https://youtu.be/gV6hP9wpMW8?t=110
[5]http://diyhpl.us/wiki/transcripts/spacex/elon-musk-making-humans-a-multiplanetary-species/
[6]https://www.nasa.gov/mission_pages/mars/news/marsmethane.html
[7]https://www.nasaspaceflight.com/2019/10/relativity-3d-printing-launch-manifest-funding/
[8]https://www.cnbc.com/2016/11/04/elon-musk-robots-will-take-your-jobs-government-will-have-to-pay-your-wage.html
[9]https://twitter.com/elonmusk/status/1091081963677437952
[10]https://youtu.be/wB3R5Xk2gTY?t=2888
[11]https://www.youtube.com/watch?v=n5ECz1YK30M
[12]http://www.skymania.com/wp/2012/04/lichen-survives-harsh-martian-setting.html/
[13]http://nextbigfuture.com/2011/07/geoneutrinos-confirm-that-half-of-heat.html
[14]http://nuclearplanet.com/Herndon's%20Nuclear%20Georeactor.html
[15]https://www.newscientist.com/article/dn11962-lab-study-indicates-mars-has-a-molten-core/
[16]https://en.wikipedia.org/wiki/Martian_dichotomy
[17]http://www.lpi.usra.edu/meetings/lpsc2011/pdf/1097.pdf
[18]https://www.theguardian.com/science/2015/may/10/crispr-genome-editing-dna-upgrade-technology-genetic-disease

Chapter 14: Who SpaceX Will Send to Mars

Phase 1 Settlers: The Hawthorne Crew

SpaceX personnel observe launch operations outside their Mission Control Center at Hawthorne Headquarters

"So who wants to go to Mars?...when you ask that question at SpaceX everyone raises their hand(1)." ~ *Gwynne Shotwell at South Summit 2015*

We should be under no illusion, sending anyone to Mars won't be easy and no doubt fraught with unknown perils. However, SpaceX has been built from the ground up to match these challenges and committed to conquering Mars. Their young engineers are remarkably adept, dedicated and relatively fearless, particularly those who have been 'home-grown', i.e. student engineers who graduated from the tough internship program to become full-fledged SpaceX engineers.

It seems more than likely a fair proportion of the first crew to land on Mars will be chosen from this peer group – who better to maintain and operate the Starship and in situ resource equipment than the people who actually built it. Due to the huge intervening distance, the communications lag between Earth and Mars ranges from 3 to 22 minutes (and sometimes even suffers communications blackout) hence the first people on Mars will need to be highly polished engineers, skilled at thinking on their feet and comfortable using their own initiative. For this reason the Phase 1 settlers will likely have a core of SpaceX engineers, chosen for their broad technical experience and strong specialty skills (reportedly SpaceX have already compiled a list of personnel who are willing to undertake Mars flights, as a preliminary step to crew selection). When the first crews arrive on Mars, their primary objective will be to set up, operate and maintain all the equipment needed to support the next group of arrivals. These Phase 2 settlers cannot be dispatched to Mars until the situation on the ground is relatively stable, with all facilities operating at nominal levels.

Some more specialist equipment, like solar arrays, power storage systems and surface transport, could be manufactured by sub-contractors. In which case specialist engineers could be recruited from these outside companies to join the first group of salient engineers. No doubt NASA would prefer to be represented amongst the first settlers but unfortunately their astronauts might fail to make the cut for this extremely select group.

"...what we should be transporting are scientists and engineers. Not pilots, really. Dragon doesn't need pilots...it should be easy to go on a spacecraft. Like, you should be able to just get on with no training and go. It shouldn't be hard(2)." ~ *Elon Musk*

NASA astronauts are normally selected and trained to work on geocentric space stations like the International Space Station (ISS). During a typical working day on-board the ISS, these astronauts have every task carefully programmed and monitored, in other words their whole day is enacted to a strict script of activities. Unfortunately, this regimen is the opposite of what will be required on Mars, where settlers will need to act independently, due to the unavoidable communication problems (lag or dropout). If, for example, a piece of life support equipment is malfunctioning, it is imperative the problem is remedied immediately by skilled personnel on site. Any attempt to engage in a delayed debate with a committee of experts back on Earth could result in dire consequences for personnel and possibly ruin any chance of salvaging vital equipment.

A good illustration might be the problems faced by the Apollo 13 astronauts when a malfunction forced them to temporarily abandon their Command Module. Unfortunately, after the three man crew retreated into the LEM (Lunar Excursion Module), they discovered it could only provide adequate life support for two crew members. At the time, it seemed entirely possible the whole crew would pass out from carbon dioxide poisoning. The air filter fitted to the Command Module was designed to support three people but agonizingly it couldn't be used on the LEM because it didn't fit the LEM's life support equipment. NASA quickly formed a group of engineers to try to overcome this problem on the ground. Fortunately, they came up with a procedure which allowed the Command Module filter to be jury rigged to fit the LEM life support equipment, effectively upgrading it to handle the extra crewmember. Step by step the engineers painstakingly relayed this jury rig procedure to the Apollo 13 crew, who were able to discuss in detail how each step should be performed and overcome any problems. This likely saved the Apollo 13 crew and clearly demonstrates the effectiveness of NASA's flight controllers and engineer support team.

However, if this same scenario had occurred close to Mars, the radio lag or dropout would have severely slowed the fault diagnosis and rectification process. Likely the crew would have all become unconscious from carbon dioxide poisoning before they could have completed the filter modification procedure. Hence those previously earthbound engineers will need to be present onboard all future Mars missions, their system skills and experience will literally become indispensable to mission safety and success. Unsurprisingly, the engineers at SpaceX already practice many of the skills necessary for Mars, i.e. personal initiative, teamwork, technical resource, creativity, determination and endurance. In other words, the conditions at SpaceX have developed the specific skill-sets and character strengths needed to cope with the challenges they will likely face on Mars.

Physical requirements are unlikely to be a problem for these comparatively young engineers, in large part they will be working in microgravity en route or reduced gravity on Mars. Under normal circumstances, performing any kind of work in a pressure suit would be extremely physically taxing because the 1 bar air pressure inside the suit makes it difficult to bend or flex. In this case SpaceX will probably opt to use one of the more advanced suit designs such as the Mechanical Counterpressure Biosuit being developed by MIT(3), which should allow almost complete freedom of action, like a second skin.

MIT Mechanical Counterpressure Biosuit

"We've learned our lessons from the Columbia accident investigation report, and one of the things that was really featured in that report was the deleterious effect of body-worn mass. One of the things we were very careful to try to eliminate as much as possible is that body-worn mass, so it's [SpaceX's spacesuit] mostly all soft goods with the exception of the helmet, and it's got a bunch of other advances that make it really usable from an operational perspective(4)." ~ *Garrett Reisman, Former SpaceX Director of Crew Operations.*

SpaceX owes a lot to NASA, so it's entirely possible some NASA personnel might be added to the crew roster for the first Mars flight. These astronauts could fill relatively high profile roles like flight specialists or pathfinder explorers, leaving the SpaceX engineers free to manage the Mars colony infrastructure. This should be equitable for both parties; NASA gains the limelight, while SpaceX achieve their Mars colony and all the promise that holds for long-term space development.

Phase 2 Settlers: Scientists, Technicians, Engineers, Explorers and Researchers

Once the Phase 1 settlement is established, SpaceX can open their doors to mainstream colonists. Space agencies from around the world will no doubt contract to have their own personnel on Mars, e.g. scientists, engineers and explorers. Large teams will need to be sent - because there's literally a whole planet to explore! Hopefully in Phase 2 they should be able to resolve whether life exists or existed on Mars. Perhaps too they will resolve why Mars's ecosphere abruptly collapsed after such a promising start. Certainly this information would be of great value to Phase 3 settlers as they attempt to terraform the planet.

Lots of novel equipment will be needed to support their efforts and grow the settlement in this alien environment, which will require some extremely creative engineers and technicians. Fabricating equipment on Mars should be possible using advanced additive manufacturing technology. Iron is found in abundance on the surface and carbon, nitrogen, oxygen and hydrogen are readily available for producing plastics. During Phase 2 a lot of effort will be focused on making in situ resources and equipment, which will no doubt require a whole army of technicians and engineers.

Perhaps an unfortunate truth but many nations and corporations will likely pay close attention to Mars settlement. No doubt some foreign intelligence agencies would give their eyeteeth to be included in Phase 2 settlement, given the sumptuous technology on display. From a spy's perspective the Mars settlement would be an all you can eat buffet of rare delicacies.

Spy is an ugly word; given their (relatively) benign behavior* these individuals could perhaps be thought of as information collectors or researchers, who work part time for an ulterior agency. If nothing else, having this demographic on-board should assist colony development through diversifying the social and genetic makeup. SpaceX was originally set-up to stimulate space activity, so engagement by other nations and corporations would, to some extent, be encouraged. A future where SpaceX entirely dominates space is the exact opposite of their intent. In the twentieth century competing nations drove space exploration, but space won't completely open up until it becomes truly commercial. So fostering commercial enterprise, from all nations, should be encouraged to help push the space frontiers.

*It's important to differentiate here between the more mundane spies who are normal people recruited or encouraged (by various means) to gather information, as opposed to professional security service operatives. Professional operatives are hardcore and quite possibly fanatical, willing to countenance anything, even sabotage at the cost of their own life. Likely these more extreme spies will be screened out during the thorough colonist selection process.

Phase 3 Settlers: The New Pilgrims

Firefly class freighter

No doubt following another hundred years of expanding government, reduced opportunities and personal liberty here on Earth, many will crave or even demand the opportunity to move to the relatively free and soon to be terraformed Mars. Prospects will certainly seem a lot brighter on the red/green planet and a broad demographic of people will want to experience this virgin new world. For practical reasons healthcare should still be free on Mars, which could encourage health tourism and retirees. The generally higher living standards will attract many types of people for a variety of reasons. For those wanting even greater adventure such as the outer planets or even interstellar travel, Mars will likely be the hub of the new space economy, hence the ideal jumping off point for such excursions. Who knows their population might even support a fair proportion of AI's per capita, which would be ironic considering Elon Musk's prominent opposition to unbridled AI development(5).

[1]https://youtu.be/omBF1P2VhRI?t=755
[2]https://youtu.be/kO1xklu-sfs?t=506
[3]http://www.universetoday.com/118939/why-cant-we-design-the-perfect-spacesuit/
[4]http://spaceflightnow.com/2016/10/13/nasa-has-no-plans-to-buy-more-soyuz-seats-and-it-may-be-too-late-anyway/
[5]http://www.theguardian.com/technology/2014/oct/27/elon-musk-artificial-intelligence-ai-biggest-existential-threat

Chapter 15: SpaceX Ideals and Mars Economy

WHAT SPACEX REPRESENT

As we've discovered, SpaceX does things a little differently to other companies. Notably, their profits are largely reinvested into the company to develop new technologies and facilities, instead of paid as premiums to shareholders or bonuses to top tier management. Employees are inspired to achieve the company's ideals and work incredibly hard, over long hours, in many cases under pretty tough conditions. This results in wartime levels of technological advancement and production efficiency – but without anyone having to be killed. During the two world wars of the twentieth century each country's inhabitants largely set aside their differences and personal agendas, even their need for luxuries, which resulted in enormous technological growth and unprecedented productivity. SpaceX have proven they can summon this vast human potential and focus it to save the human race, both physically and spiritually. Maybe because we are trapped on Earth, humanity has become increasingly introverted and mired in self interest at all levels; individual, corporate and national.

SpaceX have demonstrated there is a way to escape this quagmire by becoming more outward looking. Instead of fighting wars, they are calling us to the ultimate challenge: the struggle to traverse the limitless expanse of space and the battle to construct a new home in the vast tracts of alien wilderness. Instead of endless fighting they want to harness our enormous human potential and engineer our evolution to become a truly space faring civilization. This should shift the emphasis from material gain to scientific, technological and even spiritual enlightenment. Given SpaceX's polar opposite philosophy we can probably expect Mars culture to be wholeheartedly different to Earth's – particularly if SpaceX become the primary conduit to Mars (see Appendix 3: Mars Mission Comparison and Appendix 4: Will SpaceX Own Mars).

"I came to the conclusion that we should aspire to increase the scope and scale of human consciousness in order to better understand what questions to ask. Really, the only thing that makes sense is to strive for greater collective enlightenment." ~ *Elon Musk*

Misty sunrise over Olympus Mons

In order for the Mars colony to survive they will need to embrace the SpaceX ethic – then take it to the next level. Mars settlers will have to be incredibly resourceful and productive throughout their foreseeable future. To help illustrate here's a breakdown of some of the major sectors of the Mars economy which we could reasonably expect to emerge.

Space Habitats

From day one Mars settlers will be fighting the environment, right through to the time when they finally manage to terraform the planet. In all likelihood they will become highly adept at building surface and sub-surface habitats from in situ materials. Mars has an exceptionally low atmospheric pressure hence these habitat technologies could easily be adapted for use in space and become their primary export.

Space Healthcare

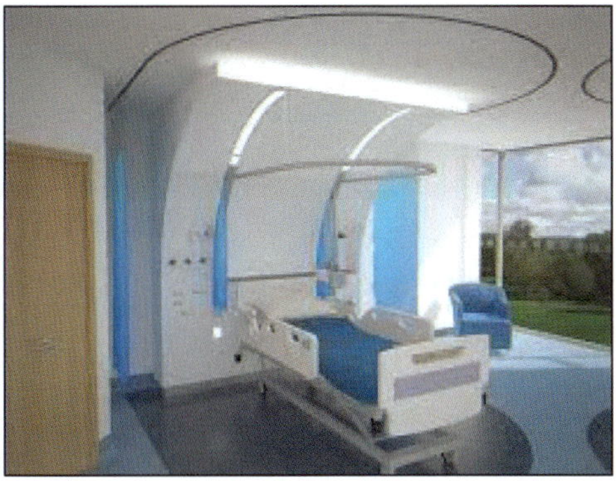

Healthcare will need to be free because it's a necessity on Mars. Everything possible has to be done to maintain settlers in peak condition in order for them to face their everyday challenges. Probably Mars physicians will need to transition to 'one shot' treatments like gene engineered cures(1) rather than lengthy and debilitating drug therapies. Given this paradigm shift, Mars healthcare could become widely perceived as the gold standard throughout the solar system and worth travelling a few extra (million) miles. Less complex cases will probably be treated remotely; the patient's symptoms and genetic profile are sent to Mars by email and the instructions for how to prepare a suitable cure sent by reply.

Mars Ethics and Education

SpaceX team at "THE AEROSPACE GAMES" (July 25, 2014)

Personal relations will become enormously important in Mars society; each person will be wholly reliant on their peers to survive. It's possible that teamwork and codes of behavior will almost become enshrined as a religion. Consequently laws might effectively become redundant, the repercussions from failing to perform as a responsible citizen would be overwhelming, a case of paradise lost. Passing on these ethics, through educational establishments, could become central to Mars economic activity. Anyone born in space will likely be sent to Mars in order for them to be raised in a reasonable gravity and to learn the Mars ethical system first hand. There would be no greater advantage you could offer a child than to be educated on Mars.

A good example of the model of education you could reasonably expect on Mars might be the school Elon Musk founded in Los Angeles, called Ad Astra(2). He wanted to improve the quality of education for his children and the children of SpaceX workers, so he established an independent school which teaches to the individual student's aptitudes and abilities. This system encourages these elementary school children to develop their problem solving skills and allows them to pursue the subjects which most inspire them at a rate and level they feel comfortable achieving. But as they say, the proof of the pudding is in the eating: -

"The kids really love going to [Ad Astra] school... they actually think vacations are too long. Like they want to go back to school." ~ *Elon Musk*

Mars Tech

Technology would likely advance at an incredible pace on Mars. Everything is new there hence some truly original thinking will be required to solve the problems associated with adapting to this alien environment. For example: energy will be the currency of survival, it's vital for life support, resource extraction, manufacturing, agriculture and synthesizing materials - like rocket propellant. Considering the amount of energy they will consume per capita, power generation technology will likely become their forte. Mars could easily become the center for nuclear research, both for power generation and nuclear drives suitable for spacecraft.

Fusion research in particular could be one area where settlers invest large amounts of resources, considering its enormous power dividend. However, the majority of Mars exports will likely be intellectual, e.g. 3D printer instructions to create Mars tech could be uploaded to wherever they're needed around the solar system.

Terraforming

An enormous amount of the economy will need to be focused on terraforming Mars. This means Mars scientists, engineers and technicians will become the go-to people for terraforming solutions because of their practical experience with a range of complementary terraforming techniques. The field could advance exponentially, likely we haven't conceived of the technologies they will eventually employ to tame their home planet.

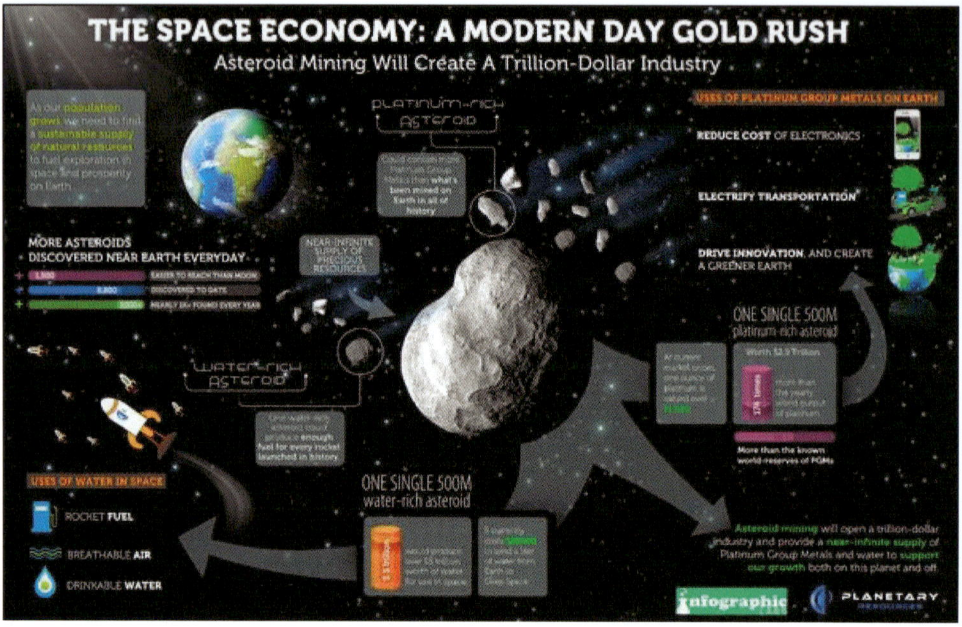

Mars too has a destiny, by the end of this century it will likely become the hub of the new space economy. The relatively low gravity means launching from Mars will be incredibly easy compared to Earth and constructing the complex hardware required for space (e.g. quantum computers and communications, AI robots etc) should be a magnitude easier there than in space. For example: early settlers will need to quickly master food production techniques and probably progress to exporting these food concentrates to mining concerns in the asteroid belt and beyond.

MARS ADMINISTRATION

Mars government could also resemble SpaceX, i.e. a relatively flat hierarchy. Ideally only a rudimentary system of government would be needed. They will probably use direct democracy to set policy and construct a minimal set of plain English laws(3). To illustrate, Elon Musk has provided some insightful views on the shape of this ideal government: -

"I think probably direct democracy is better than representative democracy, so if you are trying to represent the will of the people we are better to have direct votes which would not be possible in the old days because you had to mail things around and information moves very slowly. But in an electronic society where information moves instantly you can very directly represent the will of the people and I think that diminishes the ability of special interests to influence things in a way that is contrary to the will of the people(4)."

Sounds like lawyers and politicians could become an endangered species on Mars... No doubt the settlers will benefit from divesting themselves of many things we take for granted on Earth. It seems Mars will be a stable and secure place to live, enjoying unparalleled freedoms and technology, a true model society.

CONCLUSIONS

1. The living conditions on Mars could be relatively spartan, giving the impression that everyone was living like monks. However, settlers might argue the closeness of their community and the obvious value of their work makes them feel more fulfilled, appreciated and hopeful for the future.

2. Mars will by necessity become a technological powerhouse, possibly surpassing Earth before the end of the century. Generally, anyone who needs space tech will go to Mars.

3. If you are passionate about technology, space exploration or discovering your true potential, you might seriously consider Mars; it's getting closer every day.

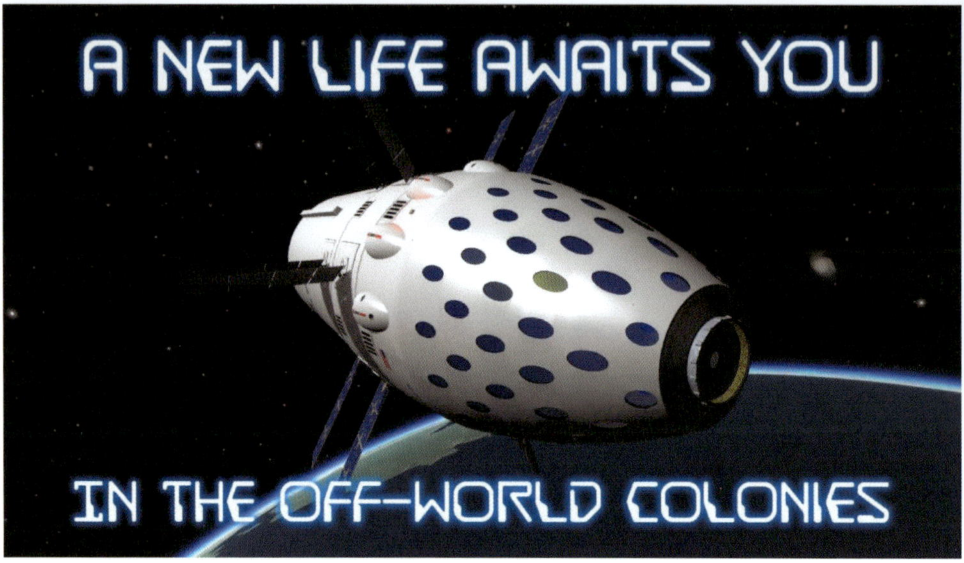

"...with the next generation of vehicles, which is going to be a sub-cooled methane/oxygen system, where the propellants are cooled to close to their freezing temperature to increase the density, we could definitely do full reusability - and that system is intended to be a fully reusable Mars transportation system. So, not merely to low Earth orbit but all the way to Mars and back, with full reusability...I think we could start to see some test flights in the five or six year time frame(5)." ~ *Elon Musk (24-10-2014)*

[1]https://www.theguardian.com/science/2016/mar/03/genetics-of-cancer-tumours-reveal-possible-treatment-revolution
[2]http://uk.businessinsider.com/elon-musk-creates-a-grade-school-2015-5?r=US&IR=T
[3]https://www.inverse.com/article/42190-elon-musk-predicts-martian-government-sxsw
[4]http://thefreethoughtproject.com/elon-musk-im-kind-pro-anarchist-talk-rebuilding-society/
[5]https://youtu.be/4DUbiCQpw_4?t=884

Chapter 16: Beyond Mars

"So what about beyond Mars...it [Starship Launch System] is actually more than a vehicle, there is obviously the rocket booster, the spaceship, the tanker and the propellant plant for in situ propellant production. If you have all four of these elements you can actually go anywhere in the solar system by planet hopping or moon hopping(1)." ~ *Elon Musk/IAC 2016*

Assuming Mars settlement is successful, SpaceX have a stretch goal to colonize the greater solar system. Starship seems well suited to this role because it can even land on worlds devoid of atmosphere, like our moon. Fortunately water and carbon dioxide are relatively abundant throughout the solar system, which should effectively provide an endless supply of fuel for long journeying spacecraft.

SpaceX propose Starship could transport a propellant plant to the outer solar system, such as the asteroid belt or one of the moons of Jupiter. This plant could then be used as a refueling station to go further, allowing them to establish a propellant depot on one of the moons of Saturn. Ideally this planet hopping process could be repeated until they have established a chain of refueling stations and colonies extending all the way to Pluto. This would open travel to anywhere in the solar system or possibly beyond into the Kuiper Belt - a worthy goal indeed.

Condensed map of the solar system, depicting planets relative size and orbits

Currently space agencies spend decades constructing specialized one shot probes to flyby planets, which produce valuable but non-definitive results. However, with a string of refueling stations Starship could land large teams of scientists on any suitable body in the solar system, producing many magnitude better results, at comparable cost. Instead of peering through the keyhole, we would be opening the door to solar exploration. And as space travel becomes more commonplace, it should create a large market for goods and materials mined or originating from these way station worlds. Likely these nexus colonies will become critical to interplanetary transport and the evolving space economy.

It should be noted that distances in the outer solar system are on a completely different scale to the inner worlds. For instance the distance between the orbits of Mars and Jupiter is 3.68 AU*, compared to Mars and Earth which is 'only' 0.52 AU(2). If Starship is to traverse these vast distances in any reasonable time, it will probably require refueling in Low Mars Orbit (LMO) or the ability to increase speed en route (possibly using in-flight refueling at way station worlds or through exploiting their superior launch characteristics). The placement and launch compatibility of these way station worlds would appear crucial, hence it might be helpful to identify the most promising candidates and explore what each has to offer.

*Astronomical Units - the distance between the sun and the earth is 1 AU or 92,955,000 miles.

THE ASTEROID BELT

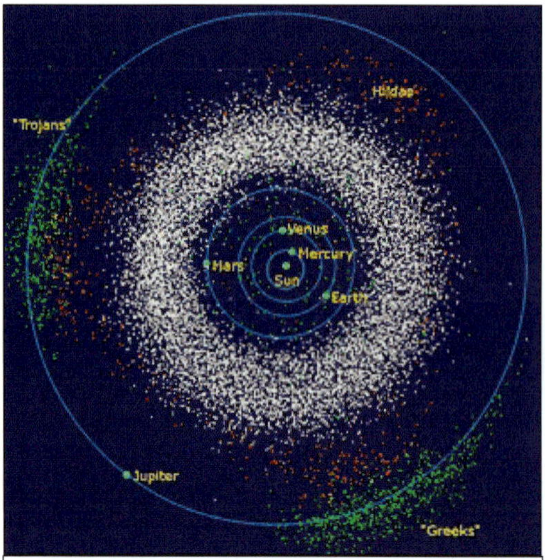

Asteroid Belt lies between the inner solar system and outer worlds

The ring of asteroids, situated between the orbits of Mars and Jupiter, would appear an unlikely place to site a refueling station, because of its proximity to Mars. However, the Asteroid Belt will probably become a hive of mining activity, due to its diverse mineral resources and relative ease of transport. All manner of noble metals and rare earths could be mined from the Asteroid Belt in quite staggering quantities. More importantly all the mundane materials required to build refueling stations, colonies and spacecraft can easily be found in the belt, making it a key resource for continued space expansion. These materials should also be less expensive to transport because of the low surface gravity on most Asteroid Belt worlds. To navigate this vast ring of asteroids, mining craft will probably require large quantities of propellant, which could also be sourced from the belt. Hence it seems likely they will need to set up multiple refueling stations in the Asteroid Belt, perhaps fairly soon after the Mars colony is established. There are many possible sites for such way station worlds, so let's try to gauge the suitability of a few of the most likely prospects.

"[What's next after Mars:] Ceres, Callisto, Ganymede & Titan(3)." ~ *Elon Musk*

Ceres (Dwarf Planet)

The dwarf planet Ceres, found just outside the asteroid belt, looks an interesting prospect for the next way station after Mars.

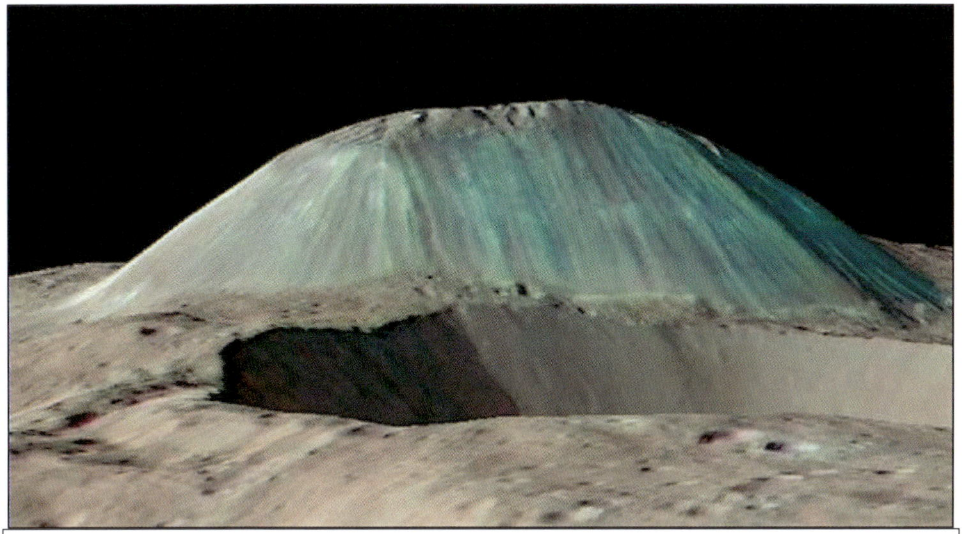

Ahuna Mons ice volcano on Ceres, credit NASA/JPL

Ceres Way Station	
Pros	**Cons**
Abundant water & carbonates for producing fuel	No atmosphere (more fuel required for landing)
Low gravity (0.029 g) requires less fuel for takeoff and landing	Low gravity could preclude long term stays or colonization
Located in Asteroid Belt (uber resources, easy to extract and transport)	Solar and cosmic radiation (no protective atmosphere or magnetosphere)
Mineral rich surface derived from asteroid impacts	No protection from asteroid impacts
9 hours 4 minute day (less fuel required for equatorial launches and landings)	High orbital inclination (10.6°from solar ecliptic plane)
Solar energy practical	Cold (-38 to -105°C)

These strong advantages likely outweigh the disadvantages, making Ceres a prime candidate to become a refueling station, both for operations in the asteroid belt and beyond. In addition, its orbital speed is different to Jupiter hence it could be used as a stepping stone for Saturn or beyond, when Jupiter is poorly aligned as a refueling stop. Spacecraft refueled and launched from Ceres could achieve much greater velocity than those originating from Earth or even Mars because of the dwarf planet's low gravity and high rate of rotation.

Alternately a Ceres based tanker rocket could rendezvous with and refuel passing spacecraft, which would remove the need to land (saving fuel and enabling increased velocity for the passing spacecraft). Cargo could also be exchanged in mid-flight, allowing Ceres Station (and any dependent asteroid belt miners) to be resupplied and enabling their mined materials to be transported on to the outer worlds or back to Mars and Earth on return flights.

Ceres overall rating: 7 out of 10 (nice place to visit...)

Vesta (Minor Planet)

Vesta Way Station	
Pros	**Cons**
Hydrated minerals & carbonates available for producing fuel	Extracting elements for propellant could be energy intensive
Low-gravity (0.025 g) requires less fuel for takeoff and landing	Low-gravity could preclude long term stays or colonization
Located in Asteroid Belt (uber resources, easy to extract and transport)	Solar and cosmic radiation (no protective atmosphere or magnetosphere)
Mineral rich surface derived from asteroid impacts	No protection from asteroid impacts
5 hour 20 minute day (less fuel required for equatorial launches and landings)	Orbital inclination (6.39°from solar ecliptic plane)
Solar energy practical	No atmosphere (more fuel required for landing)
	Cold (−60 to −130 °C)

Vesta is accompanied by a large family of lesser asteroids which could constitute a hub for asteroid mining. Unfortunately all appear fairly rocky with little evidence of surface water (vital for fuel). Hydrated minerals could supply the hydrogen and oxygen required although these minerals might be energy intensive to extract and process. Vesta is not a great place to live – unless you want to get rich!

Vesta overall rating: 6 out of 10 (it's a rock...)

Prospects for the Asteroid Belt

Depiction of Asteroid Psyche

The asteroid belt consists of protoplanet remnants and rocky materials left over from the formation of our solar system. They could be thought of as the fossilized remains of planets which failed to form during our systems primordial past. Planetary scientists could spend a lifetime exploring its dwarf planets and diverse asteroids and barely scratch the surface of all the solar genesis secrets they hold. For example, they believe the asteroid Psyche is the metallic core of a protoplanet world which was blown apart by impacts with asteroids or collisions with other protoplanets during the first billion years after the solar systems formation. Psyche is enormously dense because it consists mainly of metal, such as iron and nickel (2.27×10^{16} metric tons, i.e. a mass of 227 followed by fourteen zeros tonnes!). However, if some way could be found to descend inside this behemoth, its interior might hold more noble metals (gold, silver and platinum group metals) than have ever been mined on Earth. Literally a case of dig for gold. A combination of low gravity and easily accessible materials should prove an enormous draw for miners supplying the new space economy. In the future spacecraft and habitats will likely be built in space and there's no better place to mine such materials than the asteroid belt.

PLANET JUPITER (GAS GIANT)

ITS traverses Jupiter, credit SpaceX

Jupiter is a massive gas ball with a 'surface' gravity of 2.528 g (too high for rocket landing or takeoff) and a surface pressure that makes Venus look like a health spar. However, the moons of Jupiter look quite promising for siting a way station and later colonization. The inner moons (called the Galilean moons) appear the best prospects, unfortunately Io, Europa and Ganymede all fall inside Jupiter's fierce radiation belt, rendering them uninhabitable, unless some pretty robust radiation shielding (active and/or passive) can be engineered. This leaves Callisto as the most viable candidate because of its size and extremely low surface radiation, typically 0.1mSv/day(4). Overall the Jovian sub-system looks a great place to explore, consisting of 67 diverse moons with many more to be discovered.

Callisto (moon of Jupiter)

ITS on Callisto, credit SpaceX

Outermost of the Galilean group of moons (which orbit Jupiter) Callisto is the third largest moon in the solar system. The surface is scarred by eons of asteroid impacts with no apparent signs of tectonic motion; which means every rare mineral that has ever fallen on Callisto lies either on or near its surface - jackpot!

Callisto Way Station	
Pros	**Cons**
Abundant water and carbon dioxide for producing fuel	Minimal atmosphere (requires more fuel for landing)
Low-gravity (0.126 g) requires less fuel for takeoff and landing	Low-gravity could provide challenges for colonization
Great mineral resources derived from past asteroid impacts; more available from nearby Jovian moons	No protection from asteroid impacts
Exceptionally low radiation, (protected by Jupiter's magnetosphere)	Negligible solar energy
	Exceptionally cold (−108 to −193°C)

Callisto overall rating: 8 out of 10 (gold-digger paradise)

Prospects for Jupiter

Jupiter seems an excellent place to start our search for life in the outer solar system. The large moons Europa, Ganymede and Callisto likely have liquid oceans beneath their icy crusts, capable of supporting marine life.

Currently the innermost Galilean moons: Io, Europa and Ganymede, are inaccessible to human exploration because they orbit inside the fierce radiation belt which surrounds Jupiter. However, more advanced radiation protection for spacecraft, such as electromagnetic radiation shielding, could allow some limited settlements to be established on these large and promising worlds. If strong magnetospheres could be created on these moons that should allow them to support extensive surface colonies and produce a vibrant hub of close-knit worlds. These magnetospheres could be created through constructing vast superconducting coils on the surface powered by advanced nuclear reactors.

Alternately dynamo type magnetic fields could be generated inside each moon by changing the way their core georeactors work. J Marvin Herndon suggests that abrupt shifts in Earth's magnetic field were caused by spontaneous changes in our planet's core georeactor(5) which also suggests there could be levers for controlling georeactor activity on these large moons.

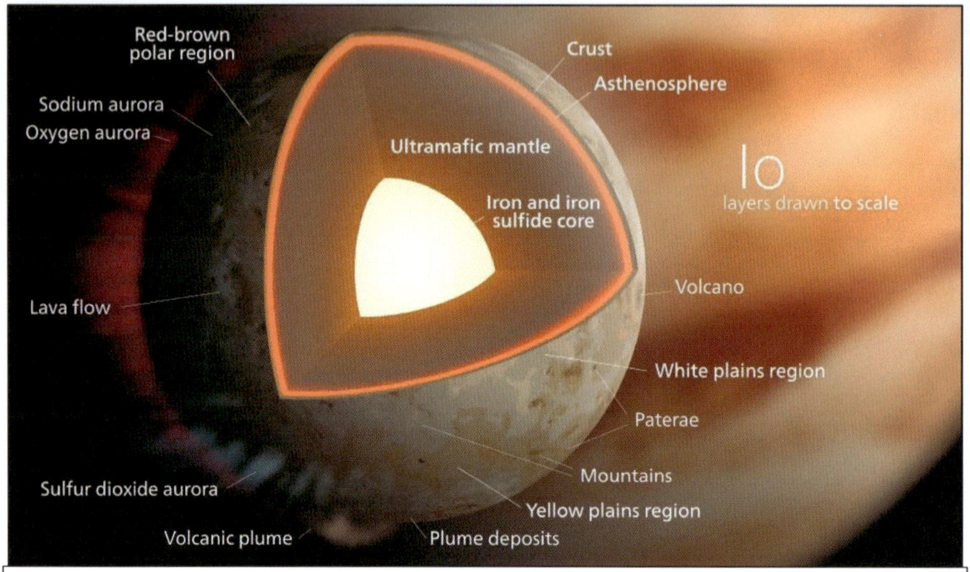

Cutaway illustration of Io, credit Kelvinsong/Wikipedia

Io's extreme geologic activity has been attributed to tidal heating from Jupiter and the other Galilean moons. However, in recent years this theory has become increasingly challenged by discoveries on Earth, Enceladus and Pluto, where volcanic activity is caused by 'inexplicable' internal heat. However, as we travel out in the solar system and explore exposed cores of protoplanets (such as the asteroid Psyche) we will likely find much of this 'inexplicable' heat is produced by spontaneous nuclear fission in their georeactor cores.

Ganymede seems an even more promising prospect because it already possesses a weak magnetosphere, which suggests there is some present or past core activity. Ganymede is an enormous moon, larger even than the planet mercury which makes it fantastically attractive as a colony world.

In all likelihood multiple techniques will be used to produce magnetospheres on such worlds, each moon will probably require a tailored approach using a different combination of many planetary engineering technologies. However, it would seem prudent to ensure the extent of any indigenous life is well documented (and understand the impact on their ecology) before permitting such a radical planetary scale intervention.

PLANET SATURN (GAS GIANT)

Starship reaches Saturn, credit SpaceX

Similar to its gas giant neighbor, Saturn has crushing air pressure and gravity making it unsuitable for conventional rocket landings. By comparison Saturn's sub-system, consisting of Enceladus, Titan and 60 smaller moons, appear positively welcoming, with much to recommend them from a resource perspective and research purpose.

Enceladus (moon of Saturn)

ITS Spacecraft on Enceladus, credit SpaceX

Enceladus is an ice world, characterized by powerful geyser jets which spray snow across the surface or eject ice crystals directly into space. This barren desolation has little to recommend it as a colony world, yet appears an ideal location for a refueling stop.

Enceladus Way Station	
Pros	**Cons**
Abundant water and methane for producing fuel	No atmosphere (requires more fuel for landing)
Low gravity (0.0113 g) requires less fuel for takeoff and landing	Low gravity could preclude long term stays or colonization
Located in Saturn system (consisting of 62 moons, some possibly habitable)	Increased geologic activity (possible hazard to habitats and installations)
32.9 hour day (reduces fuel required for equatorial launches and landings)	Exceptionally cold (−128 to −240°C) Starship would require deep space modifications
Low radiation (protected by Saturn's magnetosphere)	Negligible solar energy
	Asteroid impacts

Spacecraft will have to be modified to handle the extreme cold if they wish to visit Enceladus. For instance, the cryogenic propellant might need to be heated because oxygen freezes at −218.79 °C, methane at −182 °C. Otherwise Enceladus appears a reasonable place to site a refueling station for spacecraft visiting Saturn's inner moons. Lack of atmosphere means it is less suitable to receive interplanetary spacecraft, however, there are plenty more candidates for aerobraking in the Saturn system, including Saturn itself.

Following the Cassini mission, it has been discovered that Saturn's E ring contains a variety of organic compounds (such as methane), which are known to originate from Enceladus(6). Heat from the moon's rocky core has produced a liquid ocean beneath the icy crust, where these compounds are likely formed. Periodically when plumes of water erupt from icy fissures at the south pole, these compounds are flung into space and preserved within ice crystals, which form and replenish the E ring. Life on other worlds might use alternate compounds than the organic compounds on Earth but this discovery is very encouraging for the prospects of finding life in Enceladus' ocean. Likely it is shielded from space radiation by the thick layer of surface ice and warm enough to support indigenous life, with an abundant supply of elements percolating up from the rocky core. Even if life doesn't exist on Enceladus, it promises to be a great place to drill for methane propellant.

Overall Enceladus rating: 7 out of 10 (Enceladus cold...)

Titan (moon of Saturn)

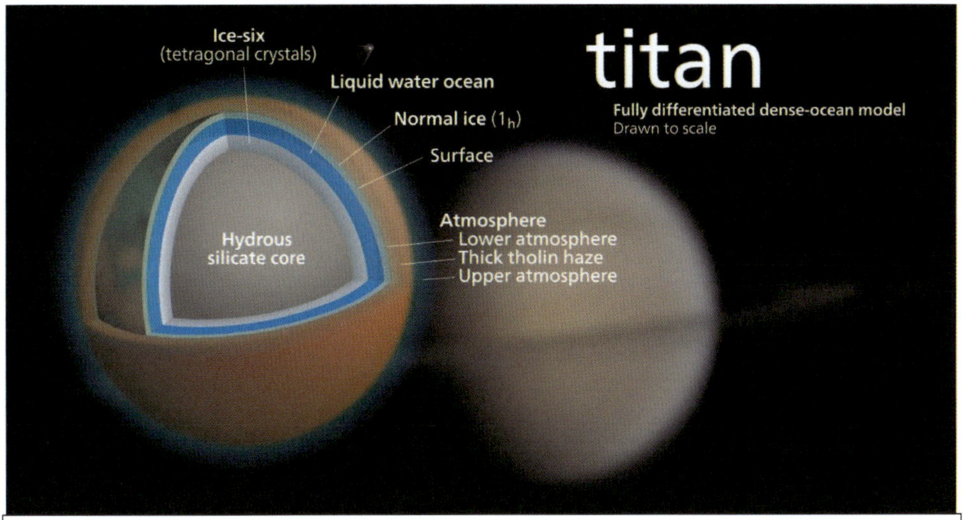

Saturn's moon Titan showing cutaway composition, credit Kelvinsong/Wikipedia

Titan could be regarded as the jewel in Saturn's crown. For a moon it is surprisingly large and possesses an unusually dense atmosphere. Atmospheric pressure is comparable to Earth's, so colonists could go out without a pressure suit (only an environment suit and oxygen mask required). Titan's orbit places it roughly in the middle of many other large explorable moons (Lapetus, Hyperion, Rhea, Dione, Tethys, Enceladus and Mimas) making it ideally situated for a base camp.

Titan Way Station	
Pros	**Cons**
Dense atmosphere reduces fuel required for landing	Non-breathable atmosphere (requires environment suit)
Reduced gravity (0.14 g) requires less fuel for takeoff and landing	Reduced gravity means remedial treatment required for continuous habitation
Located in Saturn system (consisting of 62 moons, some possibly habitable)	Negligible solar energy
Abundant minerals for building colonies/spacecraft	Some minerals might require excavation
Low radiation when inside Saturn's magnetosphere (~95% of time)	Bursts of radiation when orbit exceeds Saturn's magnetosphere (mitigated by a thick atmosphere)
Abundant methane and water for producing propellant	Extremely cold (−179.5 °C average surface temperature)
	Strong surface winds possible

The surface consists of ice boulders, interspersed with lakes of liquid methane, so no problem filling up here! This hyperabundance of hydrocarbons should allow almost limitless amounts of plastics to be produced for constructing colony structures or spacecraft. Ambient surface temperatures could be improved if some means were found to reduce atmospheric smog. Later on it might be possible to terraform Titan, making it an interesting water world.

Titan overall rating: 9 out of 10 (great place to live – except for the smog)

Prospects for Saturn

None of Saturn's moons could be described as garden spots, however, they should offer vast new territories for mining prospects, scientific investigation and terraforming research. Titan in particular has a dense and complex atmosphere and is considered one of the best places to search for life, although it is unlikely to resemble anything found on Earth.

PLANET URANUS (ICE GIANT)

Uranus is called an ice giant, which is similar in some ways to a gas giant except cryogenically cold. Pressure at the surface makes it untenable for spacecraft to land, to illustrate: the pressure on Uranus is so high and temperatures so low that it is thought to contain lakes of liquid diamond!

Similar to the gas giants, Uranus hosts numerous moons (27 known), although none of them appear to have any appreciable atmosphere. The choice then comes down to available resources to decide which moon is the most suited to become our way station world.

Titania (moon of Uranus)

Titania is Uranus's largest moon, making it a good candidate for a refueling stop, based on available raw materials and surface gravity. Although any of its sister moons might be feasible, considering they appear to mostly consist of a carbonaceous rocky core overlaid by a thick layer of surface ice.

Titania Way Station	
Pros	**Cons**
Abundant water and carbon dioxide for producing fuel	No atmosphere (requires more fuel for landing)
Low gravity (0.0386 g) requires less fuel for takeoff and landing	Low gravity could preclude long term stays or colonization
Located in the Uranus sub-system (consisting of 27 moons)	Extremely Cold (−184 to −213°C)
Asteroid debris (could require excavation)	Asteroid impacts
Low radiation (protected by Uranus's magnetosphere)	Negligible solar energy

Titania overall rating: 7 out of 10 (noble snowball)

Prospects for Uranus

Uranus has 13 known rings, of unknown composition, possibly remnants of a pulverized moon or moons. Mining these flat rings for valuable elements should be relatively straightforward because of the ease of detection and accessibility for collection. From a purely scientific perspective, researchers might be highly interested to discover why Uranus has an axial tilt of 97.77°, which means it effectively orbits the sun on its side.

PLANET NEPTUNE (ICE GIANT)

Voyager 2 image of Neptune, credit NASA/JPL

Similar to its neighbor Uranus, Neptune is an inhospitable ice giant with little prospects for colonization. However, its 14 nearby moons provide a cornucopia of possibilities for refueling, mining and exploration.

Triton (moon of Neptune)

Composite picture of Triton, credit NASA

Triton is the last of the 'big T' worlds (after Titan and Titania) and looks a great prospect for a way station world. Unusually Triton orbits retrograde (i.e. opposite to the planet's rotation) which suggests it was once a wandering dwarf planet which became snared by Neptune's gravity. This implies Triton could have a unique composition and offer materials not readily available from other moons in the Neptune sub-system. Interestingly Triton possesses a tenuous atmosphere and is geologically active, which probably provides some geothermal heating (stopping surface temperatures from falling to absolute zero!).

Triton Way Station	
Pros	**Cons**
Abundant water and some carbon for producing fuel	Tenuous atmosphere (requires more fuel for landing)
Low gravity (0.0794 g) requires less fuel for takeoff and landing	Low gravity could preclude long term stays or colonization
Located in the Neptune sub-system (consisting of 14 moons)	Increased geologic activity (possible hazard to habitats and installations)
Asteroid debris (could require excavation)	Asteroid impacts
Offers materials unique to Neptune sub-system	Extremely Cold (average −235.2 °C)
Low radiation (protected by Neptune's magnetosphere)	Negligible solar energy

Triton overall rating: 8 out of 10 (safe harbor)

Prospects for Neptune

Mostly unknown – which should be a huge draw for explorers, researchers and miners. Triton could have been captured from the Kuiper belt or possibly originates from beyond our solar system. It could easily possess extraordinary mineral resources and potential for scientific research.

"Nothing is so fatal to the progress of the human mind as to suppose that our views of science are ultimate; that there are no mysteries in nature; that our triumphs are complete, and that there are no new worlds to conquer." ~ *Humphrey Davy (chemist/philosopher)*

KUIPER BELT

The Kuiper Belt could be thought of as a second asteroid belt, located beyond Neptune's orbit, at the outer rim of the solar system (30 to 55 AU from the sun). These objects likely consist of frozen balls of gas and water with an inner core of rock. Only recently discovered, comparatively little is known about the Kuiper Belt, except for its existence.

Pluto (dwarf planet in Kuiper Belt)

Surface of Pluto

Pluto may not be the ideal choice for the terminal way station but on balance it seems likely to be the first selected. Haumea and Makemake aren't much further into the Kuiper belt (relative to the distance already travelled), however, we know a great deal more about Pluto than we do about these more exotic outlier worlds. Pluto has many moons (5 that we know of), which could provide valuable resources for building a research station or exploration craft.

Pluto Way Station	
Pros	**Cons**
Abundant carbon monoxide, and water available for producing fuel (methane mining possible)	Tenuous atmosphere (requires more fuel for landing)
Low gravity (0.063 g) requires less fuel for takeoff and landing	Low gravity could preclude long term stays or colonization
Hosts a sub-system of 5 moons	Geologically active (possible hazard to habitats and installations)
Asteroid debris (could require excavation)	Asteroid impacts
	Negligible solar energy
	Extremely Cold (-218 to -240°C)
	Highly inclined orbit (over 17°) relative to the solar ecliptic
	Cosmic Radiation

Carbon monoxide could be used to produce propellant on Pluto in place of carbon dioxide. If propellant is derived from water and CO2, more oxygen is produced than technically required to fuel spacecraft. Hence using carbon monoxide should be a more efficient way to produce propellant because it consumes less energy.

Pluto overall rating: 7 out of 10 (last stop...)

Kuiper Belt Potential

Our favorite Kuiper Belt object Pluto is a long way from Earth (between 2.66 and 4.67 billion miles) but in some regards that could be seen as an advantage. For example, interstellar voyages will probably require some form of faster than light travel to be practical, such as the Alcubierre drive. Such drives could be inherently hazardous (Alcubierre effectively exceeds the speed of light by warping space/time) hence siting a star-port in the outer solar system would seem a wise precaution. By the time we reach Pluto there is likely to be numerous hazardous undertakings which require a remote facility (e.g. superintelligent AI research, exotic matter and antimatter production etc) and there's nowhere so remote as this end of the line world. Pluto's inclined orbit raises it high above the ecliptic plane, so it could certainly be described as splendid isolation.

Planet 9 (aka Planet X)

Exploration of our solar system has barely begun and it seems possible a large undiscovered planet orbits our sun beyond the Kuiper Belt, which astronomers tentatively call 'Planet 9'(7). If this planet is proved to exist it opens the possibility that a whole series of planets might be found even further out, similar in size to the giant planets in the outer solar system; although this discovery would allow us to more correctly perceive them as mere middle solar system worlds!

BEYOND MARS CONCLUSIONS

1. Many promising sites for refueling stations (e.g. the large moons Enceladus and Titania) lack any appreciable atmosphere, which would seem to rule them out as interplanetary way stations (some atmosphere is generally required to brake from high interplanetary speed). However, it's possible Starship could use the parent planet's atmosphere to reduce velocity before landing propulsively on one of these daughter moons.

2. On many outlier worlds, solar power would be impracticable because their enormous distance from the sun makes sunlight incredibly diffuse. This suggests some form of nuclear power would be indispensable on way station worlds, to synthesis the necessary quantities of rocket fuel and help protect from the cold.

3. Adapting Starship to explore the outer solar system would require some additional work but this should be relatively minor compared to the challenges already overcome in its construction.

4. Way stations need only be built where there is local requirement e.g. for research and resource mining. For instance, if Jupiter's radiation belt is deemed too high a risk, it could easily be omitted from the chain. Starship doesn't have to stop at each way station; in fact this might add travel distance because it is rare for the planets to perfectly align (due to their differing orbital speeds around the sun).

5. Most of the candidates for way station worlds appear barren, inhospitable and unmitigatedly cold. However, these conditions could be viewed as advantages; what better place to manufacture and store cryogenic fuels than worlds where the average temperature is already at cryogenic levels and entirely uninhabited.

6. Many moons in the outer solar system possess liquid oceans, either on their surface or in their core, which might harbor life. Discovering extraterrestrial life will likely change everything, resulting in a much higher emphasis on space from civil, military and commercial.

7. It is possible that we could create an intrinsically safe nuclear reactor which could power a large propellant-less drive, e.g. an anuetronic fusion reactor(8) feeding a high performance EmDrive(9). If this could be accomplished by the time we reach Mars, it would effectively remove any need for these methalox refueling stations. However, multiple megawatts of power would likely be required to produce sufficient thrust to compete with the advanced chemical engines used by Starship. At present any chance of constructing such an ideal drive appears relatively low but the possibility cannot be discounted. As space culture rises, our future becomes increasingly fluid.

[1]https://youtu.be/H7Uyfqi_TE8?t=3763
[2]https://theplanets.org/distances-between-planets/
[3]https://twitter.com/elonmusk/status/1170983609492103168?s=21
[4]http://www.webcitation.org/5jwBSgPuV?url=http%3A%2F%2Fzimmer.csufresno.edu%2F~fringwal%2Fw08a.jup.txt
[5]http://nuclearplanet.com/prsl1994.pdf
[6]https://academic.oup.com/mnras/article/489/4/5231/5573821
[7]http://www.space.com/31670-planet-nine-solar-system-discovery.html
[8]http://lppfusion.com/
[9]http://arc.aiaa.org/doi/10.2514/1.B36120

Chapter 17: Spacefaring Civilization

So after long duress and sacrifice we finally arrive at the promised land and transform into a spacefaring civilization. But what does that mean precisely? Considering the profound impact on our species and culture, the term 'spacefaring' is certainly something worth unpacking. Perhaps the best way to illustrate how these advances will affect us is through drawing parallels to similar periods of advancement in the past. History tends to go in cycles, so one of our past cycles of development could form a good analogy for what we might conceivably expect in the future.

In the second millennium we experienced three distinct phases of development, beginning with the Renaissance, followed by the Age of Discovery and concluding with the Industrial Revolution. This progression occurred naturally, building on advances made in the previous phase, in a seemingly inevitable cascade of development. Assuming history is cyclic, that might suggest we are entering a Second Renaissance era. It's inherently difficult to evaluate any time while it is still evolving so let's compare what we know about the first renaissance with our current stage of development.

RENAISSANCE

In fourteenth century Italy, many individuals started to question their medieval beliefs and began to experiment through creative art, detailed investigation of natural biology including human anatomy, imaginative speculation and innovative machinery - plus a whole lot more. There are many reasons why this all started in Florence, such as: -

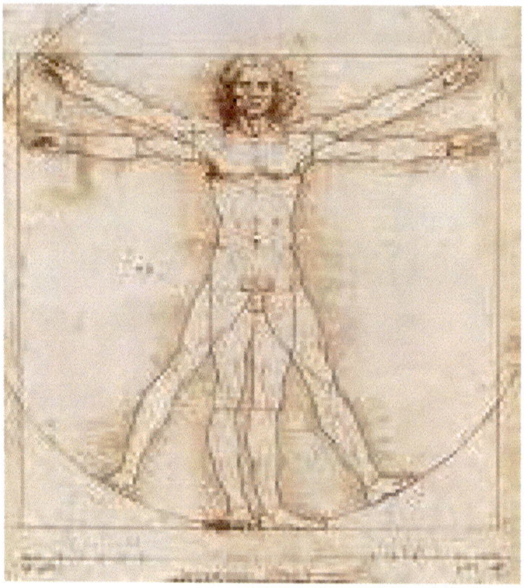

- a critical mass of talented people
- more personal freedom
- availability of wealthy patrons
- influx of ancient Greek manuscripts (aka 'lost works') after the fall of Constantinople
- improved communication of ideas via printed books

Thanks to freethinkers like Leonardo da Vinci, the renaissance proved that small teams of talented individuals, who work tirelessly to investigate and innovate, are the ideal way to produce advances and expand the realm of human understanding. Their successes opened new areas of activity, allowing severe regulations from local ordinance, trade guilds, feudal and papal control to be circumvented, which led to a far more enlightened era. Antique attitudes that only powerful nobility, merchant houses and the church could substantially change things for the better were steadily eroded as a new approach called 'science' began to prove its worth.

In the modern era we see close parallels to innovative approaches which began in Silicon Valley and were then replicated around the world. The Valley certainly achieved a critical mass of talented people for computer science, in what is arguably the most liberal part of the free world. Venture capitalists (i.e. wealthy patrons) helped these talented individuals explore their ideas through funding startups - the modern equivalent to Leonardo's workshop. In addition communication of ideas has steadily improved due to the internet, which Silicon Valley arguably made possible through its groundbreaking work on semiconductors and ARPANET (the first incarnation of the internet). Previously written information was locked away in myriad libraries, now almost everything is available at our fingertips, even translated from foreign languages, all thanks to the internet.

Whereas the original renaissance expanded and embellished our world, the work begun in Silicon Valley has augmented our reality and created new virtual worlds to explore. Aided by the internet, mobile technology and the new freedoms they allow,

Silicon Valley advances make almost anything seem possible. SpaceX is a prime example of a second generation startup (along with Tesla, SolarCity, Amazon etc) which can trace its roots back to this Silicon Valley renaissance.

Such creativity cannot be restricted or constrained and could arguably lead to a paradigm shift in public perception. For example: until recently it was perceived only governments could perform nuclear fusion research, make meaningful improvements to the environment or attempt space programs. But citizens have stepped up and are making a difference through private fusion companies, installing solar panels/energy storage and of course the commercialization of space.

Citizen Scientists at LPP Fusion with Dense Plasma Focus Reactor

What might previously have been regarded as dubious science, such as EmDrive or Alcubierre research, are now being actively pursued by a range of people from citizen scientists through to NASA(1), who has arguably become infected by this unbridled enthusiasm. Opening our minds and believing in our potential has allowed many new possibilities to flourish and should now admit humanity to the next level of development.

AGE OF DISCOVERY

Following the original renaissance new ideas flooded Europe and people were willing to take more risk, like challenge common misunderstandings (the world is flat, our sun circles the earth etc) and conservative church dogma. Following the fall of the Byzantine Empire, trade with Asia became difficult and consequently Eastern goods, like spices and silks, increased in value. Some new trade routes were opened up, although the Far East remained tantalizingly out of reach. Enter Christopher Columbus, a Genoese Italian, who following the advice of the Florentine astronomer Paolo dal Pozzo Toscanelli, attempted something quite original. He proposed to take a shortcut to Asia across the Atlantic Ocean, as an alternative to the long and perilous route around the Cape of Good Hope and Horn of Africa.

This idea was as brave as it was flawed but the risk paid off when he discovered the Americas. The discovery of two new continents sparked even more ambitious expeditions, like the voyage of Ferdinand Magellan, who managed to find a route beyond the Americas and across the Pacific Ocean to East Asia. By necessity the shipbuilders needed to develop larger oceangoing vessels, better suited to these long and arduous sea voyages. Consequently global trade routes were established allowing new and exotic goods to circulate the world, beginning the process of globalization. More importantly natural science improved by leaps and bounds as scientists (like Charles Darwin) were now free to examine and contrast all of Earth's diverse corners – leading to some startling advances. What began as a madcap adventure led to tremendous gains in commerce and science, which resulted in a stepped improvement in the wealth of participating nations.

The modern day parallels are striking. In 2002 Elon Musk, following advice from Dr Robert Zubrin, established a startup with an end goal to colonize Mars. Funded through venture capital friends and his personal fortune, SpaceX now seem poised to embark on the greatest voyage of discovery. Colonizing Mars could

certainly be seen as a close analogue to the process of founding the New World. Of necessity the journey will require more advanced spacecraft than currently available, which could then be used to take us anywhere in the solar system. The highly capable people sent to new worlds will have to establish an extensive economy to furnish all the colonist's needs, e.g. air, food, water, transport, power, heat, entertainment etc. In addition, these colonies will likely be in a state of continuous expansion, to cater for new arrivals and indigenous births. Any trade of physical goods between our worlds could be relatively minor, however, there would likely be an enormous exchange of virtual goods, such as proprietary, patented, copyrighted materials, currency, deeds and holdings. These exchanges could supercharge such trade on Earth, helping us transition into a more virtual economy. The answers to problems we have barely considered will flood Earth and our refinements will return to the outer worlds, enriching us both.

From a science perspective, each new world we explore will tell its tale about how planets are born, struggle to survive and eventually flourish or fail. This study of planetary evolution should suggest new techniques for planetary engineering, allowing us to moderate climate change on Earth and adapt the environment on suitable worlds, like Mars. At present we can only theorize about such processes, based on a single data point i.e. Earth, but full exploration should allow us to know what makes planet's tick, knowledge which will become increasingly valuable as our home planet and solar system matures.

Perhaps the most enticing prospect for this new age is the discovery of life on other worlds. Organic molecules (i.e. carbon based compounds) are fairly ubiquitous in space, which means the potential for life is everywhere. Nature has a way, through trial and error, of discovering new designs and processes; no doubt any life we find on alien worlds will exhibit profoundly different approaches to survival through exploiting natural laws. One can almost imagine the next Charles Darwin of the outer worlds and all the things they might discover. Certainly tracing the evolution of life beyond Earth will be a mighty triumph for science, all made possible by the advent of affordable space transport. Of course the discovery of new life could lead to enormous advances here on Earth, for example bringing the humble potato to Europe fed billions, arguably allowed Earth's population to double. With such riches raining from the heavens, perhaps the greatest advance will be cultural. Risk takers, visionaries and explorers will become ideals, while discovering the true potential of life our highest goal.

Finding what's out there and inside of us

INDUSTRIAL REVOLUTION

Which leads us to the final phase in the cycle: the Industrial Revolution. Classically this phase began in what became known as Great Britain, an island kingdom which used renaissance advances and the age of discovery to become a world trade hub. Faced with increased demand, labor shortages and no shortage of ideas, some enterprising textile manufacturers developed a series of increasingly complex machines to turn seemingly insoluble problems into a priceless opportunity. Starting with the most basic cotton processing gin they improved every part of the textile manufacturing process, culminating in enormously productive steam powered mills. When steam engines became available, it changed everything, allowing the development of machine tools, trains and steel ships which could power from continent to continent - finally free from the vagaries of trade winds and sail. More importantly this transition to mechanization allowed important cultural advances, such as the dissolution of the slave trade. Forced labor was no longer required to pull oars, till the earth or harvest crops which allowed more liberal attitudes to prevail. Overnight Great Britain changed from a bulwark of slavery to an active disruptor of the trade - which accelerated the benefits they derived from mechanization.

Again, Mars technological advances will likely start small then filter through the system until they become ubiquitous. The process will appear almost organic, with colonists applying lessons learned from the equipment supplied by Earth. Mars colonists will have to produce almost everything where it's needed to be truly self sufficient, including fresh approach technologies. Hence Mars appears the most likely place for this technological revolution to begin, for many reasons – ending in necessity. Fortunately these Mars techniques, like advanced additive manufacture and in situ resource utilization, could easily be adapted for Earth consumption. For instance: need a new SUV? Just drive your old car into the shop then watch as the latest Mars tech transforms it into a fashionable and economic Mars rover.

Concept Mars Rover, credit NASA

Here's a few major advances we might reasonably expect to see from this second industrial revolution: -

- **Planetary Engineering** – terraforming techniques which could be adapted to control climate change here on Earth

"In solving for a good Mars climate, we will learn a great deal about how to do so on Earth." ~ *Elon Musk*

- **Quantum Energy** – a step improvement in power generation and space propulsion

- **Quantum Communications** – instantaneous data transfer via quantum entanglement

- **High Genetics** – mastery of genetic techniques which allow us to treat all known disease. Generally less technology implanted for health reasons because organs, joints and tendons can be regenerated

- **Space Habitats** – formation of independent space colonies with self supporting economies

- **Space Liners** – behemoth spacecraft which transport goods, people and materials between deep space habitats and mega-terminals in planetary orbit

- **Direct Democracy** – direct representation becomes ubiquitous throughout the solar system, strengthening democratic process

- **Meta-humans** – extreme genetic enhancements possible, through biotechnology (e.g. ability to work in vacuum)

- **Synthetic Food** – food produced solely through artificial biologic processes

- **Superintelligent AI** – a construct which comprehensively exceeds human capabilities (which could provoke the next cycle of renaissance-discovery-revolution in the interstellar sphere)

Mars will likely be the epicenter of these changes due to local necessity and because it plugs into the expanding space economy, which promises even greater reward - both virtual and real. Anything brought from the outer-worlds will find Mars a very attractive destination, due to its low gravity, vibrant economy and secure communications link to Earth. Again, the parallels between this Mars colony and Great Britain are striking; both use the preceding period of renaissance and age of discovery to become trade hubs primed to grab the technological initiative.

Ubiquitous automation should literally free everyone from wage slavery, allowing us to pursue our passions and realize our human potential. Likely anyone who chooses to follow a profession should be even more effective and skilled because such service is voluntary, in an area which inspires them and they wish to improve.

Generally any Earth nation which helps to establish a Mars colony should reap unprecedented rewards, similar to the trading nations in the original age of discovery – and dear old Grande Bretagne.

What starts with our faith in human potential should end in a bright and enviable future. Space will become open to all, not just the privileged few. Undoubtedly space will become the ultimate stage for exploring our humanity.

CONCLUSIONS

1. Renaissance freethinkers chose to act unilaterally, resulting in sustained technical advance and substantial economic progress. Similarly, Silicon Valley type entrepreneurs (Musk, Bezos, Jobs etc) ignored the standard approach of established industries to create novel technologies and new realms of commerce. Hence strong parallels can be drawn between these two eras and we do appear to be experiencing a new technical renaissance. The original renaissance gains came about through creativity and science whereas the current age is exploring all that's possible through unfettered creativity and computer science.

2. The Age of Discovery was literally mind expanding, it opened our eyes to a larger world, allowing us to explore new cultures, technologies and commerce. Founding a new world, which will create itself through technologic advances like terraforming, direct democracy and genetic engineering, should provoke even more profound advances in the human condition. The greater the test the more we discover.

3. The Industrial Revolution resulted from the demands of the time and local conditions. Mars colonization will likely create even more pressing needs and concentrated conditions to provoke a similar technical revolution. If necessity is the mother of invention then well placed technologists are the father.

4. The progression between stages of development (Renaissance → Age of Discovery → Industrial Revolution) was precarious and not certain to succeed. For instance, if Christopher Columbus had failed to find a sponsor or failed to return from the Americas, it's quite possible the transition to the Age of Discovery would have stalled. Similarly Mars landings will need to succeed to evoke a new Age of Discovery and subsequent technological revolution.

5. All stages of development required a change in attitude regarding what is possible, with disdain for the 'cult of the expert' and status quo. Change isn't inevitable but it is necessary for us to grow.

Succinctly put: This grand enterprise we have embarked upon will be the making of humanity.

[1]https://www.nasa.gov/ames/ocs/2014-summer-series/harold-white

Appendix 1: Regulatory Hurdles and Morality Issues

OUTER SPACE TREATY (1967)

Dark lineae could indicate presence of water in this 'fresh' impact crater in the Sirenum Fossae region of Mars

SpaceX will have to overcome many challenges to reach Mars, quite possibly including international law. Their mission architecture requires landing at one of the 'Special Regions' on Mars, i.e. a water rich location which could conceivably harbor indigenous life. Such a landing could contravene Article IX of the 1967 Outer Space Treaty: -

"...States* Parties to the Treaty shall pursue studies of outer space, including the moon and other celestial bodies, and conduct exploration of them so as to avoid their harmful contamination and also adverse changes in the environment of the earth resulting from the introduction of extraterrestrial matter and, where necessary, shall adopt appropriate measures for this purpose..."

However, SpaceX are unlikely to be bound by this provision because the treaty is aspirational, i.e. sets ideals of behavior in space for national governments. If the contemporary situation changes, national governments are obligated to alter how this law is applied. A good example would be the contemporary U.S. bill: "H.R.1508 - Space Resource Exploration and Utilization Act of 2015," which allows property rights for asteroid resources under US law. This specifically amends the application of Article II of the Outer Space Treaty: -

"...the moon and other celestial bodies, is not subject to national* appropriation by claim of sovereignty, by means of use or occupation, or by any other means. However, the State that launches a space object retains jurisdiction and control over that object."

In other words H.R.1508 specifically extends ownership from the spacecraft to anything it can obtain from an asteroid. It could be argued that if a spacecraft managed to obtain a complete asteroid, the entity which controls that spacecraft would thereby gain property rights over the entire asteroid.

Deep Space Industries Harvester 2 captures an asteroid

This sets precedence for the exemption of private entities from the Outer Space Treaty. Given the considerable advantages derived from a manned landing on Mars it's possible the application of OST will be further amended to accommodate such efforts, at the relevant time. It seems likely that SpaceX, given their rate of development, will have grown immeasurably in the 2020-2025 timeframe and exert considerable influence as the prime mover of the burgeoning space economy.

*These article clauses use the terms 'states' or 'national' which suggests they might not apply to a commercial company. However, companies are incorporated under national law and hence act as agents and answerable to their host government. For example: any attempt to use a company to appropriate a celestial body would be classed as an action sanctioned by that nation. The Article II statement 'by any other means' is clearly broad enough to include companies.

As a final note, the Outer Space Treaty was ratified by the US in 1967, two years before they commenced lunar landings. During the Apollo missions thousands of pounds of equipment was left on the lunar surface and hundreds of pounds of lunar regolith was returned to Earth. Certain measures were taken to avoid harmful contamination but some cross-contamination was inevitable. This demonstrates no matter how tight the wording might appear on the OST, at the end of the day the treaty is aspirational, with no legal weight inside signatory nations. If SpaceX choose to colonize Mars there is no means to prosecute them under the Outer Space Treaty, which is international law.

"Corporations exist by virtue of being incorporated under a national legal system. Their vessels – whether ships, airplanes or spacecraft – are registered under those laws and operate as extensions of national territory when they venture abroad." ~ *Washington Post*

OST CONCLUSION

The Outer Space Treaty needs to be respected except where its strict application would proscribe space exploration.

Moon landings began two years after signing OST

MORALITY OF MARS COLONISATION

Mars and Earth formed in the same era, from similar space materials and exposed to comparable space conditions. As you might expect, early Mars was not too dissimilar to Earth and probably possessed an ocean(1), which many regard as the cradle of life (at least judging by Earth's early history). However, after a billion years or more of parallel development our histories diverge. For some reason the geological activity on Mars all but ceased and the planet effectively died. Its surface waters receded to the poles or underground and even escaped into space. Hence if life still exists on Mars it is likely to be relatively simple i.e. microbes – complex life couldn't survive the extreme rigors of current surface conditions (low atmosphere, no liquid water, high radiation etc). Finding even simple life on Mars would be enormously important but, on balance, such a discovery is unlikely to occur without manned exploration.

Viking 1 - first Mars lander

The original Viking robotic landers detected what they believed to be microbial life but because there was no one present to observe or repeat the test, this apparent positive result couldn't be verified. Which presents a dilemma, we can go to Mars to discover whether life exists but risk contaminating it, or we can assume life exists and never travel to Mars (or any celestial body) to avoid any possibility of contamination. Assuming something to be the case (i.e. that life might be present) then applying that rule universally is very poor scientific practice. Admittedly we run the risk of contaminating Mars if we choose to colonize, but in reality that means we will risk exposing Martian microbes to spores and microbes from Earth. So the morality question boils down to: "will no-one think about the microbes..."

However, it seems entirely possible that Earth contamination has already occurred, which would make the morality question moot. A few years ago dark streaks were discovered on the surface of Mars, which NASA later confirmed to be streams of seasonally occurring surface water(2). Water appears to be essential to life and certainly life is ubiquitous wherever open water is found here on Earth. Assuming the same holds true for Mars, these seasonal streams would certainly seem the best place to look for indigenous life. Unfortunately Cassie Conley (NASA's Planetary Protection Officer) subsequently discovered their Mars Curiosity Rover had passed within feet of one of these probable streams, which means contamination with Earth microbes could be entirely possible.

During the Curiosity Rover's manufacture, efforts were made to reduce the number of terrestrial microbes it harbored, however, these measures could only reduce not eliminate the rovers microbial load. Any increase in local humidity would likely be interpreted by such microbes as a green light to reproduce and so perhaps they have preceded us in building the first colony on Mars.

Fortunately any possibility of back contamination to Earth can be virtually ruled out. Anything alive on Mars would probably find our Earth environment to be anathema, our abundance of highly reactive oxygen alone would likely prove fatal. Anything which has evolved to survive on the frigid wastes of Mars will find our 'hot rock' lifestyle extremely challenging.

It should be noted that a great deal of material has already been transferred from Mars to Earth. Asteroid impacts on Mars have blasted material into space for billions of years. Inevitably some of this debris travelled through space to its closest neighbor and descended on Earth as meteorites. It's possible that any such Mars microbes have failed to survive the transit, so far, but could manage this feat at some point in the future. It would hence seem wise to take this opportunity to travel to Mars and investigate these microbes, in isolation. Even ostriches have concluded that hiding your head in the sand isn't a legitimate survival tactic(3).

Impact by a 10km asteroid caused the Cretaceous–Paleogene extinction event, in other words dinosaurs became extinct because they lacked a space program!

Notably there have already been five great extinction events on Earth, caused by radical climate change, asteroid impacts and the like. Hence restricting human activity to Earth would mean dooming us as a species. For instance, Jupiter's immense gravity field is continuously 'planing off' asteroids from the asteroid belt and hurling them into the inner solar system. However, if we choose to live and work in space, we will become more outward looking - and more likely to see the asteroids coming.

MORALITY OF MARS CONCLUSIONS

1. If complex life existed on early Mars, it is very unlikely to have survived their global climate catastrophe. If only microbes remain that effectively removes the morality issue associated with colonizing Mars.

Bones of Mars – layered rock indicates there was once a great deal of geologic activity on Mars

2. The morality of colonizing Mars is undeniable because it reduces the chance that man will become extinct should a similar catastrophe befall Earth. In sum: inhabiting two worlds should double our chance of survival.

[1]http://www.nytimes.com/2015/03/06/science/mars-had-an-ocean-scientists-say-pointing-to-new-data.html?_r=0

[2]https://www.wired.com/2016/08/shouldnt-go-mars-might-decimate-martians/

[3]http://www.livescience.com/27433-ostriches.html

Appendix 2: Mars Flight Plan

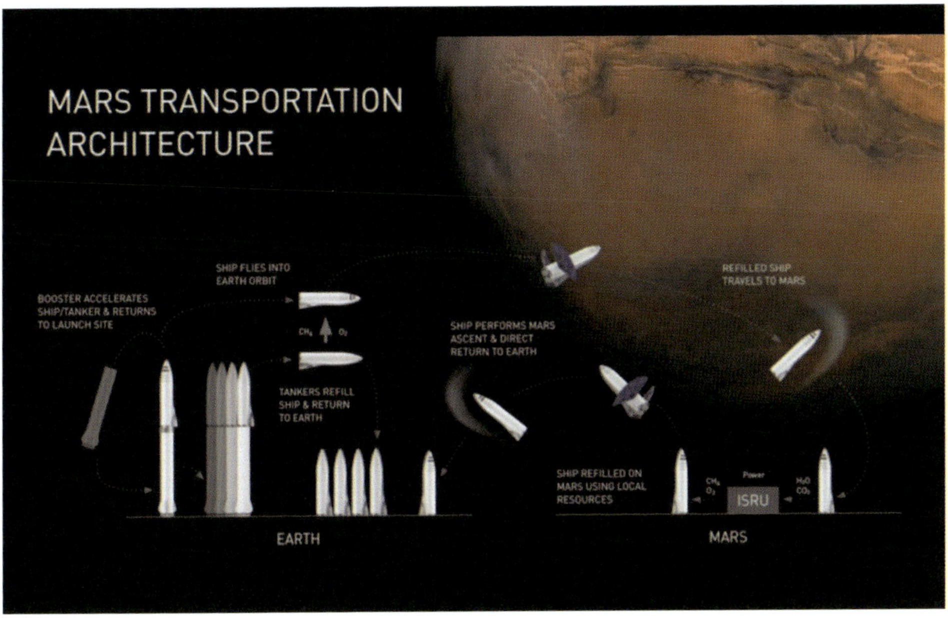

Mars Flight Plan		
Step	**Operation**	**Description**
1.	Earth Launch	Starship LS launches in a prograde inclination from a site near the equator (to maximize payload delivered to orbit)
2.	Starship Booster Reuse	Super Heavy boosts Starship into space then Returns To Launch Site (RTLS)
3.	Starship Reuse	Starship spacecraft expends nearly all propellant to reach Low Earth Orbit then is refueled by tanker spacecraft launched by Super Heavy booster
4.	Mars Flight	Starship spacecraft uses a high velocity transfer to transit to Mars. A combination of aerobraking and supersonic retropropulsion in the Martian atmosphere are used to reduce velocity, allowing Starship to safely land
5.	Mars Operations	Starship spacecraft offloads cargo and passengers then refuels using in situ resources. This allows the spacecraft to return to Earth without the need for staging or further refueling operations
6.	Starship Returns to Earth	Starship spacecraft returns to land on Earth, enabling it to be refurbished, refueled and reintegrated with the booster in time for the next planetary conjunction window

Appendix 3: Mars Mission Comparison

SLS Block 2 crew/cargo configuration and preparing to launch

Given SpaceX and NASA are both aiming for Mars, it might be helpful to compare their different approaches to solving the same problem. It seems likely such comparisons between SpaceX's Starship Launch System and NASA's SLS will become hotly debated as these parallel development programs progress towards completion. The minimum requirements to perform a Mars mission are detailed in the comparison table below.

SLS – Starship Launch System Comparison Table

Mars Mission Requirements	SLS (Block 2B)	Starship LS
Launch Period	2030-2033(1)	2022-2024(2)
Payload to LEO	130 mt(3) expendable	235 mt reusable, including Starship(2)
Payload to Mars	75 mt Mars lander to high Mars orbit, using two SLS vehicles(1)	120 mt spacecraft carrying 100 mt of cargo to Mars surface
Number of Crew	4 using Orion capsule(4)	12(5)
Trans-Mars Flight Duration	7 months(1)	3-6 months(2)
Mars Exploration Duration	2 weeks on surface, 18 months in Mars orbit(1)	20-23 months on surface (minimum stay*)
Total Mission Duration	32 months	29-32 months
Minimum Number of Launches	5 SLS + 2 commercial heavy lift vehicles(1)	10 (Crew Starship, Cargo Starship and 8 Tanker launches)
Minimum Number of Launch Vehicles	7(1)	1
Launch Cadence	1 - 2 per year(1)	48 hours(6)
Launch Campaign	3 years	2 years

SLS – Starship Launch System Comparison Table (continued)

Mars Mission Requirements	SLS (Block 2B)	Starship LS
Mission Specific Stages	Orion Spacecraft, SEP (Solar Electric Propulsion) Tug, DSH (Deep Space Habitat), EDS (Earth Departure Stage), 2 EUS (Exploration Upper Stage), TEI (Trans Earth Injection), MOI (Mars Orbit Injection), Crew Lander plus MAV (Mars Ascent Vehicle), MAV Transfer Stage(1)	Starship (crew variant), Starship (cargo variant), Orbital Tanker
Launch Vehicle Cost	$1bn(7) (does not include Orion Spacecraft and mission specific stages)	$760 million(8) (includes Booster, Crew Starship, Cargo Starship & Tanker)
Mars Mission Cost	$6 billion, plus cost for mission specific stages, e.g. $1bn per Orion(9)	$62 million(8) (assumes multiple missions using same vehicle)
Overall Cost	$100 billion(10)	$5 billion(11)
Cost to Colonize Mars	n/a	$100 billion, mainly self-funded(12)

*Note: 20-23 months is the minimum exploration time for Starship missions. If explorers wish to stay longer, this should be possible at the Mars colony settlement.

CONCLUSIONS

1. Starship LS is the most capable launch vehicle for Mars exploration due to its mission specific architecture i.e. high launch cadence, large payload, multi-reusability, on orbit and Mars refueling capabilities.

2. The SLS is more suited to cislunar missions due to its lower payload capability and low launch cadence, stemming from an expendable design. SLS payload to Mars is insufficient to support the necessary long stay missions, with a flight rate of one or two launches per year.

3. Due to its unsustainability, SLS is unsuitable for long term space exploration. It can't provide the sustained stimulus required to produce a new space economy and ubiquitous space exploration. If used to land on Mars, SLS could set the high-water mark for space exploration in the twenty first century, similar to the Apollo Moon landings in the twentieth.

4. SpaceX's Starship should provide a commercially viable means to achieve Mars colonization. This bold transplant of humanity promises to transform us into a truly space faring civilization.

'Before and after' Mars at entrance to SpaceX Headquarters

[1]https://youtu.be/BZ6B9yrcQN4?t=486
[2]https://www.reddit.com/r/spacex/comments/73cw1u/my_notestranscript_elons_iac_2017_talk_parts_1_2/
[3]http://www.nasa.gov/press/2014/august/nasa-completes-key-review-of-world-s-most-powerful-rocket-in-support-of-journey-to
[4]http://www.nasa.gov/sites/default/files/atoms/files/fs-2014-08-004-jsc-orion_quickfacts-web.pdf
[5]https://www.reddit.com/r/spacex/comments/590wi9/i_am_elon_musk_ask_me_anything_about_becoming_a/d94txm0/?context=3
[6]https://www.reddit.com/r/spacex/comments/3z0scl/spacex_job_postings_show_goal_of_48_hour/
[7]http://spacenews.com/the-big-changes-that-may-not-be-coming-to-nasa/
[8]https://www.youtube.com/watch?v=H7Uyfqi_TE8&t=43m15s
[9]https://arstechnica.com/science/2016/08/how-much-will-sls-and-orion-cost-to-fly-finally-some-answers/
[10]https://arstechnica.com/science/2017/03/new-report-nasa-spends-72-cents-of-every-sls-dollar-on-overhead-costs/
[11]https://youtu.be/zu7WJD8vpAQ?t=5272
[12]https://twitter.com/elonmusk/status/1159965968379985920

Appendix 4: Will SpaceX Own Mars

SpaceX Tee Shirt

Setting aside such philosophical questions as whether you can truly own anything, it would perhaps be helpful to define what it means to own something, like a planet. Mars won't fit in your pocket, you can't move it (much) but you could apply restrictions on access, laws on its populace, tariffs on imported goods etc, which means you would effectively own it. Hence if SpaceX become the prime mover on Mars, it could be argued they will actually own Mars, at least by applying this narrow definition. According to their Mars project timeline they should put boots on the ground at least eight years ahead of NASA's proposed SLS missions (see Appendix 3: Mars Mission Comparison). Because of their technological head start, SpaceX will likely have a monopoly on Mars transport for some time. So, if SpaceX control access, establish the colony's laws and set the price for imports, a good case could be made they de facto control Mars (if not de jure).

It seems like SpaceX could slip through the OST net because they are not a nation, merely a private corporation. However, SpaceX is certainly a US company, incorporated in, protected by and subject to their nations laws. Hence, if SpaceX did appear to be 'hogging' Mars a strong case could be made at the UN that the US was attempting to appropriate a celestial body by "other means." UN members might accept ownership of asteroids by private concerns but the appropriation of an entire planet is unlikely to be overlooked; it's a question of scale. So SpaceX probably won't have their Mars cake and eat it, in such a scenario they might even have to accept some kind of UN mission there instead. Something along the lines of: "persistent rumors of unreasonable living conditions" or "unacceptable work practices" needing investigation.

It seems the people who live, work and die on Mars might have the final say. Article II proscribes national annexation by anyone on Earth but the inhabitants of Mars would undoubtedly have a solid claim to the planet (de facto and de jure), particularly if some of them were actually born there. SpaceX should be relatively happy with this arrangement, which should allow them to distance themselves from their difficult relationship with Mars. At the end of the day, SpaceX are not into land grabbing, all they want to be is the engineer on a big locomotive, i.e. they only want to be a space transport company.

"Mars belongs to the Martians(1)." ~ *Elon Musk*

So if Mars chooses to declare independence and the only company who can reach them decides to cooperate, this will probably happen quietly, quite possibly sooner rather than later. Settlers certainly won't sit by while 'externals' (SpaceX or anyone else on Earth) attempt to make their lives harder through unnecessary taxes, influence or regulation.

The early Mars settlers will be of majority US extraction and no doubt vocal should they perceive any such erosion of their freedom. Mars natives might even reject UN membership because they're "too Earth" - a case of: "Mars can take care of its own problems." Undoubtedly there are some interesting times ahead – and sooner than you might think.

Falcon 9 launch and landing (long exposure),
SpaceX don't own space but they've certainly made their mark on it

[1]https://twitter.com/elonmusk/status/1142892702968631296

Appendix 5: Meta Conclusions

We've explored a lot of new territory and flown quite quickly over some interesting terrain features. Probably it would be helpful to review some overall conclusions from this journey.

1. The SpaceX plan for manned Mars exploration, while ambitious, appears practical given the inherent difficulties. Their paradigm step in space technology perhaps allows us some insight into why manned exploration of Mars has not been previously possible.

2. The ISS has provided a large stimulus for space development. Building a city on Mars will likely complete the process, producing a space economy which will eventually exceed Earth's economic activity.

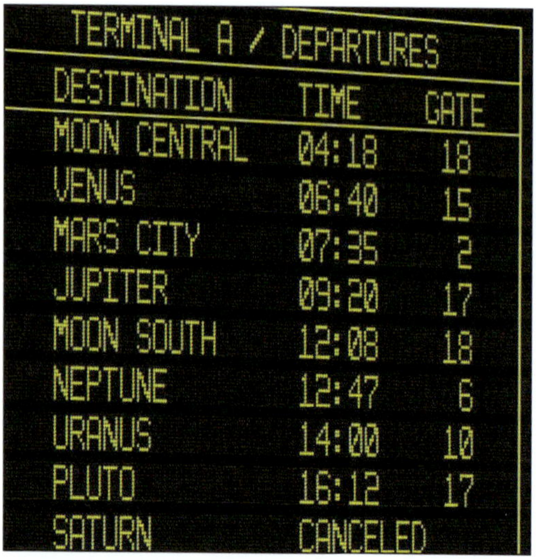

3. The development of space travel is somewhat analogous to the development of air travel, e.g. the first aircraft were relatively small, cheap and unreliable, progressing through to bigger more reliable and rapidly reusable airliners. Originally all air development was amateur then, during the First World War, it transitioned to the military. When the Great War ended commercial concerns adapted the wartime technology gains to create the first air travel industry. Similarly rocketry originated with amateurs, was advanced by the military/NASA during the Cold War and now private enterprise will likely use this same technology to fully commercialize space.

"Five years ago, this would have been like, 'No way, what are we doing asking commercial providers to be able to do this?' Now it feels like a natural progression for space travel(1)."
~ *Sunita Williams, veteran NASA Astronaut*

4. NASA is a great space agency but it's unreasonable to expect them to operate a space passenger service, construct a city on Mars or build asteroid mining infrastructure; such endeavors are best accomplished through commercial enterprise. However, NASA could easily become a full partner with SpaceX in order to facilitate Mars colonization. Such a transition should take place in the next few years after Starship demonstrates its sustainable approach to space exploration.

5. Perhaps Elon Musk's greatest accomplishment is to have inspired people to change the world and give them the opportunity to realize their dream. Mostly they have done this work with joy, for the greater good of humanity – instead of feeding human greed.

6. SpaceX intend to deploy Starship (a spacecraft with comparable pressurized volume to the ISS) by the end of 2020 – something truly remarkable. This should allow construction of a moon base, which would be amazing, and a colony on Mars – utterly stupendous. But achieving these goals isn't as important as the passion SpaceX have brought to space development and demonstrating it can be a practical commercial enterprise. The space genie is out the bottle.

7. Human history has changed. Fated to die on a small planet in an outer spiral arm of the galaxy, we now grasp destiny with both hands and begin our journey of discovery amongst the stars. No doubt this seminal act will be celebrated through the millennia.

As we draw to a close I feel it is only fair to allow you-know-who the final word: -

"Now is the first time in the history of Earth that the window is open, where it's possible for us to extend life to another planet. That window may be open for a long time – and hopefully it is – but it also may be open for a short time. I think the wise move is to make life multiplanetary while we can(2)." ~ *Elon Musk*

[1]http://uk.businessinsider.com/nasa-astronaut-suni-williams-on-spacex-boeing-spaceships-2018-6?utm_source=reddit.com
[2]https://www.space.com/31388-elon-musk-colonize-mars-now.html

Glossary

AI – Artificial Intelligence, a computer program with human-like or better cognitive faculties

AGI – Artificial General Intelligence, computer program which performs at human levels of intellectual ability

ASI – Artificial Super Intelligence, computer program which exceeds human intellectual ability

ARLM – All Roads Lead to Mars, meaning everything SpaceX does either directly or indirectly assists their goal of achieving Mars colonization

AU – Astronomical Unit, the distance between the sun and earth is 1 AU or 92,955,000 miles

BC – SpaceX launch facility at Boca Chica Texas

CCDev2 – Commercial Crew Development (part 2), NASA's program to assist the development of commercial crew launches to Low Earth Orbit

Cislunar – lying between Earth and the moon, including lunar orbit

DSG – Deep Space Gateway, original name for the space station NASA intend to build near the moon

DSN – Deep Space Network, radio antennae equispaced around the world, used to track and control spacecraft

DST – Deep Space Transport, the long range spacecraft NASA intends to use for exploration missions beyond lunar orbit

ECLSS – Environmental Control and Life Support System, the equipment which maintains adequate conditions for habitation (e.g. reasonable temperature, breathable air and potable water) onboard a spacecraft or space station

EDL – Entry, Descent and Landing, the deceleration process used by spacecraft from atmospheric entry through to landing

F9 – Falcon 9, a Heavy Lift launch vehicle built and operated by SpaceX

FH – Falcon Heavy, a Super Heavy Lift launch vehicle built and operated by SpaceX

Glossary (page 2)

FSD – Full Self Driving, system being developed by Tesla to allow their cars to operate without drivers

GEO – Geostationary Earth Orbit, an orbit where the angular speed is equal to that of Earth's rotation, making the satellite appear to hover over a fixed geographical location

GTO – Geostationary Transfer Orbit, an intermediate elliptical orbit used when launching geostationary satellites

IAC – International Astronautical Congress, an astronautics and space engineering conference

ICBM – Inter Continental Ballistic Missile, a heavy lift launch vehicle, primarily designed to deliver nuclear ordinance

IoT – Internet of Things, physical devices, vehicles and appliances which can be monitored and controlled via the internet

Isp – Specific impulse, a measure of the efficiency of rocket and jet engines

ISRU – In Situ Resource Utilization, i.e. using locally sourced materials

ISS – International Space Station

ITS – Interplanetary Transport System, an early version of Starship Launch System

Kerolox – a kerosene and liquid oxygen system

LAS – Launch Abort System, the emergency system designed to thrust a crew capsule away from a failing rocket

LEO – Low Earth Orbit

LMO – Low Mars Orbit

LOP-G – Lunar Orbital Platform-Gateway, the current name for the space station NASA intend to build near the moon

LZ – Landing Zone, a landing area for rocket stages

Glossary (page 3)

MCB – Mechanical Counterpressure Biosuit, developed by Massachusetts Institute of Technology

Methalox – a methane and liquid oxygen system

MCT – Mars Colonial Transporter, an early name for Starship Launch System

MIT – Massachusetts Institute of Technology

OST – Outer Space Treaty, the UN agreed framework for space law

RTLS – Return To Launch Site

SHL – Super Heavy Lift, a launch vehicle capable of lifting 50 metric tons (110,000 lb) to LEO

SLS – Space Launch System, the SHL launch vehicle NASA intend to use for deep space missions (principally SLS Block 1 and Block 2)

SST – Super Sonic Transport, a supersonic aircraft designed to carry civilian airline passengers

SSTO – Single Stage To Orbit, a launch vehicle capable of reaching orbit under its own power i.e. without using a booster stage

Starship LS – Starship Launch System, i.e. the full stack launch vehicle comprised of the Starship upper stage and Super Heavy booster.

USSR – United Soviet Socialist Republics, a group of soviet republics, led by Russia, between 1922 - 1991

VLEO – Very Low Earth Orbit

Chris Prophet Bibliography

Synopsis

An inordinate encounter spawns a parlous friendship. Ana Salikova, Israeli Msc student, and 'Gabriel,' a larger than life enigma with alien social skills, are drawn into fathomless intrigue. Primeval forces conspire to erase a quantum leap in human evolution. The environment appears a forlorn hope - unless a back door can be found to a Euphoric Eden. Ineluctable secrets threaten to cleave their mismatched alliance. Beyond expectation they engender a startling strategy for survival - yet nothing escapes their avid eyed demons. If you're seeking innovative entertaining escape, stop when you reach 'Euphoria.'

Links

Prophet's third novel: "Euphoria" is currently published by Amazon for Kindle.

USA Link: http://www.amazon.com/dp/B00E8K6NG8

UK Link: http://www.amazon.co.uk/dp/B00E8K6NG8

Synopsis

Marooned amongst the pyroclasm wrought by Euphoria, Gabriel faces his final and most momentous adventure. Only stark choices remain: die honorably for his clan or explore the twisting paths of darkness, hazarding his soul.

The fate of Euphoria's monumental hero entwines with the destruction of a hidden world. Wracked by guilt he chooses the path of the true Champion. Lady Luck maybe radiant but is she fair?

Links

The Euphoria sequel novel: "Euphoria Exsolution" is currently published by Amazon for Kindle.

USA Link: http://www.amazon.com/dp/B00GMO4ANQ

UK Link: http://www.amazon.co.uk/dp/B00GMO4ANQ

Synopsis

Starpoint Military Academy is shocked by the news our solar system will be visited by the rogue planet Omnis. Three cadets, who barely tolerate one another, are selected for an extraordinary exploratory mission; Xia Green a tech gifted metahuman, Troy Alexander a natural leader from Mars and Corrine Marcus beguiling heir to the vast Ares Corp Empire.

The visiting exoplanet becomes a devil's playground when they find themselves stranded with dubious allies. Taking destiny in their hands they forge a new path for the Allied Worlds, while combating external and inner terrors. Earth might barely survive the choices they make – but Omnis is a brave new world.

Link

"Corrine's Star" is currently published by Anderian Designs Ltd on DriveThruFiction.

https://www.drivethrufiction.com/product/224660/Corrines-Star

Printed in Great Britain
by Amazon

36278305R00130